G. I. Taylor was one of the great physical scientists of this century, and was notable for the originality and independence of his thinking. He made outstanding contributions to the mechanics of fluid and solid materials, and his ideas have had wide application in meteorology, oceanography, mechanical, civil and chemical engineering, hydraulics and materials science. He was both an experimenter and a theoretician of distinction, and saw instinctively the simplest approach to the investigation of a new phenomenon. He was involved in the early development of meteorology, aeronautics and metal physics, and was consulted widely on the problems in mechanics arising in the two World Wars. He was a keen small-boat sailor, and helped himself and thousands of others throughout the world by his brilliantly imaginative new design of an anchor. As a man he was gentle and lovable, with needle-sharp wits which hurt no-one..

How was Taylor able to be so innovative? This interesting and unusual mix of science and biography helps us to answer that question. The author was a graduate student of Taylor and a friend for 30 years, and is well placed to describe his achievements and his life, most of which was spent in Cambridge. He does so without introducing mathematical details, making this book understandable and enjoyable, especially for those whose own interests have brought them into contact with the scientific legacy of G. I. Taylor.

The Life and Legacy of G. I. Taylor

Frontispiece: photo of oil painting of Geoffrey Taylor by Ruskin Spear, 1966

The Life and Legacy of G. I. Taylor

George Batchelor, FRS

*Emeritus Professor of Applied Mathematics in
the University of Cambridge*

CAMBRIDGE
UNIVERSITY PRESS

CAMBRIDGE UNIVERSITY PRESS
Cambridge, New York, Melbourne, Madrid, Cape Town, Singapore, São Paulo

Cambridge University Press
The Edinburgh Building, Cambridge CB2 8RU, UK

Published in the United States of America by Cambridge University Press, New York

www.cambridge.org
Information on this title: www.cambridge.org/9780521461214

© Cambridge University Press 1996

First published 1996
This digitally printed version 2008

A catalogue record for this publication is available from the British Library

ISBN 978-0-521-46121-4 hardback
ISBN 978-0-521-00231-8 paperback

Contents

Contents

Preface

I have wanted for some years to write a book about G. I. Taylor, who died in 1975. He was one of the great physical scientists of this century, and was notable for the originality and independence of his thinking. He made outstanding contributions to the mechanics of fluids and solids and their applications, and pioneered several important fields in mechanics. He was both an experimenter and a theoretician of distinction. On the personal side Taylor generated great admiration and affection among those he met. I believe that the physical-science community would be enriched by more knowledge of what he accomplished in mechanics, and also of how he did it, for this too is part of his legacy. Since I am greatly indebted to Taylor for his friendship and for his influence on my own research life, I felt I owed it to him to write his biography. I saw this book as a posthumous tribute to a great man, as a source of insight stimulated by Taylor's work, and as a picture on paper of a truly simple, good, man. And writing the book has been a labour of love for me.

So great is my respect for Taylor that it seems audacious to make value judgements about him. Even so, I have tried to avoid making this book simply a long eulogy. I have not hesitated to point out what I believed to be a weakness of character or of – rarely – a scientific development. I admire him greatly but I hope that has not rendered me incapable of seeking the truth.

I should present here my credentials as a biographer of Taylor. So far as the technicalities of writing are concerned, I have written previously two scientific books, one a research monograph and one a textbook for students, and a number of papers on different aspects of fluid mechanics, and I believe that the requirements of a biography are not very different from those of a scientific book. In both spheres one tries to organize and present the material in a way which is clear, precise and – if possible – elegant. The essential difference is that the biography contains a human element whereas scientific writing normally does not. My experience of compositions containing a human element is not zero, but is limited to a few obituary memoirs, the longest being a memoir on G.I. Taylor for the Royal Society in 1976[1]. This present biography of Taylor can be regarded as, roughly speaking, an enlarged version of the Royal Society memoir, and the reception given to that memoir encourages me to think that this larger-scale description of both his life and his work is on the right lines. I take this opportunity to thank the Royal Society for permitting me to incorporate extracts from my Royal Society memoir of G.I. Taylor in the present book.

Readers will wish also to know whether I knew Taylor well and whether I have had access to the existing papers about Taylor and his life. The short answer to these queries is a definite 'yes'. I first met Taylor when I arrived in England in April 1945. He was 59 at the time, I was 25, and he and I were both in Cambridge for the remaining 30 years of his life. He was scientifically active during that period, being only a little less prolific in his output than in the years before 1945. He supervised my work for a PhD during the period 1945–48, and as a friend later I was aware of what he was doing scientifically. During the period 1958–1971 I edited and Cambridge University Press published in four volumes *The Scientific Papers of Sir Geoffrey Ingram Taylor*. When he died in 1975 he left the paper records of his life and work in a disordered state in one room of his house. At the request of Miss Gladys Davies, who lived in the house with Taylor, and to whom he left the house on his death, I sorted Taylor's papers roughly, and learnt a great deal about the man while doing so. The documents and records and correspondence regarded as being of some interest were then arranged and classified by officers of the Contemporary Scientific Archives Centre to form a collection which the Council of Trinity College agreed could be deposited in the College Library. The Librarian has kindly allowed me easy access to the papers in the collection for the purpose of writing this book. The collection is extremely valuable, although it suffers from the

1. *Biographical Memoirs of Fellows of the Royal Society*, vol. 22, 1976, pp. 565–633.

drawback that it is by no means complete and in particular contains relatively few letters written by Taylor since he usually wrote letters by hand without keeping copies.

The author of a biography of a scientist faces a difficult question of selection of material. At what level of scientific education should the scientific developments described in the book be understandable? The problem is especially acute in the case of a physical scientist like Taylor whose work involves mathematical relations and ideas. Going deeply into his papers would have made the book both long and hard-going for the non-specialist reader. I therefore adopted the working rule that my description of Taylor's scientific work should be concise and generally intelligible to a graduate student who has had some introduction to fluid or solid mechanics. To this end I have avoided all mathematical manipulations and have concentrated on conveying a qualitative understanding of the more important investigations made by Taylor. I have tried to reveal the physical ideas involved and to show the nature of his reasoning and the ingenuity of his experiments. Some readers may find my explanations too brief, and there will be experienced readers who find my explanations too superficial. However, a reader whose interest in one of Taylor's publications is stirred will be able to consult the appropriate reference in the complete bibliography of Taylor's works at the end of this book.

The magnitude of Taylor's total scientific output also demands some selection of the papers to be described in this book. Volume I of his *Scientific Papers* contains 41 previously published papers on aspects of the mechanics of solid materials, and Volumes II, III and IV together contain 152 papers on fluid mechanics. It is not feasible to describe them all, and I have chosen for brief description the more significant papers, in particular those which are representative of a group.

The relatively small representation of papers on the mechanics of solid materials in this book reflects my limited competence in that field. I am greatly indebted to my colleagues Rodney Hill and Nevill Mott for being willing to contribute descriptions of Taylor's very important work on plasticity of crystalline materials to chapter 11. As a consequence of these two authoritative contributions, from the continuum and the microscopic (or dislocation) viewpoints respectively, and the reproduction of a historical note by Taylor on the early stages of dislocation theory, chapter 11 gives a clear and interesting account of Taylor's research on plasticity. However there are relatively few references to Taylor's work on mechanics of solids outside chapter 11, which I regret.

In his later years, Taylor wrote a number of 'popular' articles and texts of addresses which I found to be attractive and informative additions to his scientific works. Their subjects include his gifted family, his experiences at the Royal Aircraft Factory helping to lay the scientific foundations of aeronautics, his participation in the *Scotia* expedition, his sailing exploits, his travels in remote parts of Indonesia, the history of some important scientific developments in which he was involved, and his personal philosophy of research; and many are unpublished. All these articles are clear and entertaining, and they tell us a good deal about the man. There would be some loss of insight if I tried to condense them, and I concluded that the most satisfactory plan would be to reproduce several of the articles in full, at appropriate places in the text. If the book consequently seems to some readers to be a quasi-autobiography, so be it; I have no doubt that readers will enjoy Taylor's popular writing and will find it rewarding.

Works by Taylor in the complete bibliography at the end of this book will be referred to in the text simply by the year of publication or production, with the addition of letters a, b, c ... to distinguish between works produced in the same year; for example (1954d). The further designation in the bibliographical list, SP IV, 34 for example, indicates that a paper has been reprinted in volume IV of Taylor's *Scientific Papers* and is paper number 34 within that volume. References to publications by other authors are usually given in footnotes to the text.

Many people have helped me to produce this book about Geoffrey Taylor and his work. Some have helped by being encouraging, some have provided information about G.I., some have helped me to understand pieces of his research, and some have provided helpful comments on my drafts. It has been my good fortune that everyone approved of the proposal to write a book about G.I., and everyone was willing to contribute in some way. I have of course benefitted from the universal admiration and affection in which the memory of G.I. is held. Many times I have reflected that my task would have been much more difficult, and less pleasant, if my subject had been a lesser man.

In the early days when I was uncertain about the structure of the proposed book I received valuable comments and advice from Philip Drazin, Tim Pedley, and David Tranah of CUP.

Sir Arnold Hall, Rodney Hill, the late Harry Jones, Sir Nevill Mott, and the late Lord Penney all knew G.I. and his work, and at my request kindly contributed descriptions of areas of G.I.'s research unfamiliar to me. These contributions were incorporated first in the Royal Society Biographical Memoir I wrote in 1976 and now again, with the consent of the Royal Society, in the present biography.

Other people who told me what they knew about G.I. are Albert Green, Sir William Hawthorne, Bertha Jeffreys and Dick Scorer.

I thank John Stewart for making for me a translation into English of Prandtl's letters to Taylor in the twenties and thirties, my secretary Karen Stringer for coping patiently with the endless revisions, and my wife Wilma for delving into the records concerning the foundling who was given the name Taylor and became G.I.'s grandfather on his father's side.

There were many draft descriptions of particular aspects of G.I.'s life or work on which I needed a second opinion, and people who gave me valuable comments, sometimes concerning a paragraph or two and sometimes about whole chapters, were Grisha Barenblatt, Barbara Garlick, Julian Hunt, Anne Keynes, Frans Nieustadt, and John Willis. I owe a particular debt of gratitude to Keith Moffatt who generously undertook to read a complete draft and whose comments were both encouraging and critical.

G.I. Taylor: Chronology

22 April 1885	Marriage of Margaret Boole and Edward Ingram Taylor
7 March 1886	Birth of Geoffrey Ingram Taylor in London
1897–1899	Attended University College Preparatory School at Holly Hill
1899–1905	Attended University College School, in Gower Street, London
1905–1908	Studied mathematics and physics at Trinity College, Cambridge
1908	Awarded a major scholarship by Trinity College and began research at Cambridge on a problem suggested by J.J. Thomson
1910	Published first paper on fluid mechanics, on the structure of shock waves, for which he was awarded a Smith's Prize at Cambridge
October 1910	Elected to a Prize Fellowship at Trinity College
April–July, 1911	In Linford Sanatorium, Ringwood, with pleurisy
1 January 1912	Appointed to a Readership in Dynamical Meteorology at the University of Cambridge for three years
1913	Served as a meteorologist on an expedition, on the sailing ship *Scotia*, to report on icebergs in the North Atlantic, following the sinking of the *Titanic*
August 1914	Joined a group of civilians at Farnborough helping the Royal Flying Corps to put the design and operation of aeroplanes on a scientific basis
1915	Awarded the Adams Prize for the period 1913–14 at University of Cambridge for an essay on 'Turbulent motion in fluids'

1917	Appointed meteorological advisor to the Royal Flying Corps
1918	Advisor on meteorology and navigation to the Handley Page group trying to be the first to fly across the Atlantic
1919	Elected to Fellowship of the Royal Society
October 1919	Appointed to a Fellowship and Lectureship in Mathematics at Trinity College, Cambridge
1923	Appointed to a Royal Society Research Professorship at Cambridge. Walter Thompson becomes Taylor's technician
15 August 1925	Married Stephanie Ravenhill, a school teacher in Birmingham
1927	Cruise to the Lofoton Islands on *Frolic* with Stephanie
1929	Participation in 4th Pacific Science Congress in Java. Tours in Borneo and Japan with Stephanie
1933	Awarded a Royal Medal of the Royal Society
1944	Knighted. Awarded the Copley Medal of the Royal Society
16 July 1945	Witnessed the first test of an atomic bomb in New Mexico
1946	Awarded the US Medal for Merit
1951	Retired from Yarrow Research Professorship
4 June 1967	Stephanie Taylor died
July 1969	Appointed to the Order of Merit
April 1972	A severe stroke and some loss of mobility
27 June 1975	A second and fatal stroke at Farmfield

CHAPTER I

An introduction to G.I. Taylor

G eoffrey Ingram Taylor was born on 7 March 1886 in St John's Wood, London, and died at the age of 89 in Cambridge after having lived there for most of his life. He was one of the most notable scientists of this century, and over a period of more than 60 years produced a steady stream of research papers of the highest originality. He occupied a leading place in applied mathematics, in classical physics and in engineering science, and was equally at home and equally respected in these three disciplines. Taylor's work is of the greatest importance to the mechanics of fluids and solids and to their application in meteorology, oceanography, aeronautics, hydraulics, metal physics, mechanical engineering and chemical engineering. His research in fluid mechanics in particular could be said to have given the subject much of its present character. He stands in the great British tradition represented by Kelvin, Rayleigh, Reynolds, Richardson and Stokes, although he got more from experiments than any one of these men – and he was in my judgement the most original. He had the rare honour of seeing his scientific papers, some previously unpublished, gathered and published in four thick volumes during his lifetime. Taylor was elected to Fellowship of the Royal Society in 1919, knighted in 1944, awarded the US Medal for Merit in 1946, and appointed to the Order of Merit in 1969.

He was also an adventurer. In 1913 he served as meteorologist in an expedition on the old sailing ship *Scotia* sent to get information about icebergs on the Newfoundland Banks following the disastrous sinking of the *Titanic*, and seized the opportunity to measure the vertical distribution of wind velocity, temperature, and humidity over the sea surface by lifting instruments up to heights of about 2500 metres by means of kites and balloons. In the following year he joined a group of civilians at the Royal Aircraft Factory at Farnborough helping the Royal Flying Corps to put the design and operation of air-craft on a scientific basis, and learnt to fly in order to have direct experience of an aeroplane as a mechanism and, later, to parachute. He was among the first British people to enjoy the sport of skiing in Switzerland; he was a keen rock-climber; and he tried at the age of 80 to master water-skiing. In 1929 he and his wife, Stephanie, explored on foot remote parts of Borneo where the natives had not previously seen a white person.

But sailing was his greatest love. As a schoolboy he designed and built a boat in his bedroom, and sailed on the Thames, sleeping on board overnight. In later years he owned larger sailing boats, in parti-cular *Frolic*, a cutter 48 feet long in which he and Stephanie made some enterprising voyages, including a cruise up the coast of Norway to the Lofoton Islands for which he was awarded the Royal Cruising Club Cup for 1927. In all these activities the spice of adven-ture was mixed with intellectual curiosity about how things worked and how improvements might be made. Among the improvements was a revolutionary design of anchor for small boats.

Geoffrey Taylor was a member of a distinguished family. Scientific creativity of high order had appeared in several of his ancestors, most notably in his grandfather, George Boole, who founded the study of mathematical logic.

Behind Taylor's immense scientific achievements and adventurous pursuits was a modest, gentle, lovable man with a razor-sharp mind which was never used to hurt anyone. He had the engaging curiosity of a bright child, and retained that fresh enquiring attitude through-out his life, even into his eighties. He was rational in his approach to everything, and confident of his abilities. His childhood in a stable secure family unit was a happy one, and his marriage, although child-less, was successful. This serene background to his personal life as a child and as an adult no doubt helped him to develop and retain a well-balanced contented uncomplicated personality. He was also extremely independent, both in his personal life and in his scientific work. Present-day scientists will be impressed to learn that, to the best of my knowledge, he never had a secretary, never took leave away from Cambridge (except to advise on problems of national importance), and never applied for a research grant.

Taylor had the gift of naturalness, which enabled him to face any task or problem, scientific or non-scientific, without stress or self-concern. He knew what he wanted to do with his life, namely, to study mechanical phenomena. Some perceptive institutions awarded him the means to do exactly that; and he proceeded to do it supremely well. He had every reason to be happy, and he was. The degree of naturalness that Taylor possessed is uncommon among members of the human race, and I believe it is the primary source of his outstanding success as a scientist. Most great men have wide interests and varied experiences, and they develop a complex network of interactions with other people. By contrast, Taylor was a simple man whose mind was uncluttered and free at all times for the scientific enquiry that he loved above all else. His character and his activities were perfectly matched, and the self-regard and irrationalities and self-made obstacles to success which beset most of us did not exist for him.

There is a paradox here, in that the life of such a simple straightforward man may seem not to justify publication of a biography. There might seem to be little needing to be said beyond a description of his scientific achievements. But Taylor's research contributions were only half the story. His simplicity of character and outlook was a source of great scientific strength, and we should enquire how he was able to make simplicity such a positive quality. He left us a legacy, which was characterized at an international symposium at Cambridge in March 1986 to commemorate the 100th anniversary of his birth as 'fluid mechanics in the spirit of G.I. Taylor'. This legacy contains a lesson for scientists, in particular for those involved in the training of young people for research.

He summed up his attitude to life with characteristic modesty and light-heartedness at the end of a lecture of reminiscences (1952e); 'One never regrets the things one has done. It is the things one has not done when the opportunity came that cause the bitterest pangs.' It seems unlikely that Taylor had cause for many pangs of regret, for in another lecture given in the same year he said: 'I think that if I were to start again I should still try to be an applied mathematician, because the number of amusing activities to which mathematics can lead one is so great.'

CHAPTER 2

Taylor's family

George Boole

Taylor's mother came from an interesting and talented family, and he was proud to be a part of it. The family tree shown in figure 2.1 begins with John Boole, Taylor's great-grandfather on his mother's side, who was a cobbler in Lincoln. John Boole had a lively mind and wide interests which included the making of scientific instruments and observing the stars with his own telescope. He was a poor man, and his eldest son, George, who was born in 1815, felt obliged to go to work at an early age despite an intense wish to be educated. George's formal education went little further than elementary school, but after some help with mathematics and classical and modern languages from his father and family friends he made very rapid progress by his own efforts. Difficult to credit though it may be, the cobbler John Boole and his home-educated son corresponded with each other in Greek. George was drawn increasingly to mathematics and, while still in his teens and working as a school teacher, read Newton's *Principia* and Lagrange's *Mécanique Analytique*. He began independent mathematical work, and became especially interested in the laws of combination of symbols representing mathematical operations. His profound originality was soon evident. George Boole was awarded a Royal Medal by the Royal Society in 1844, at the age of 29, for his contributions to symbolic logic, and was elected a Fellow in 1857. His opportunities to pursue his mathematical research increased greatly when in 1849 he was appointed the first Professor of Mathematics at the newly founded Queen's College, Cork. It was here that he met and married Mary Everest, with whom he had five children, all girls, before his death in 1864.

Much more information about George Boole and his work than I have space for here may be found in the recent biographical study by MacHale.[1]

Taylor wrote several engaging articles about the life of George Boole, two essentially the same on the occasion of the centenary of the publication of Boole's *The Laws of Thought* (1954f, 1956e) and a third for a meeting in Lincoln to mark the centenary of Boole's death (1964c), which Taylor attended as the only survivor of Boole's eight grandchildren. Taylor brings out well the almost saintly character of his grandfather. These articles are clearly and attractively written, like others by Taylor in his later years, and they tell us things about him as well as about Boole. I am therefore reproducing here the greater part of Taylor's address on the life of George Boole to the Royal Irish Academy (1954f), with the incorporation of some short extracts from the later version (1956e):

> I have been asked to give a short account of the life but not the scientific works of my grandfather. For this task I fear I am little better fitted than any other student, for Boole died when my mother, his second daughter, was only six and I have met only two people who remembered him. One was my grandmother, who survived her husband for 52 years.[2] Fortunately in 1878 she wrote sketches of her life with Boole in the now defunct *University Magazine* and these give a clear picture of his short married life from 1855 until his death in 1864.[3] They reveal an attractive and sensitive person whose thoughts and actions were much affected by a religion which he felt deeply but seldom mentioned.
>
> Boole came from a family of farmers and small tradesmen who lived in and round Lincoln. The earliest member of the family who has been traced is Joshua Boole, who was born in 1670. None of the family seems to have been in any way remarkable except John Boole, George Boole's father. John Boole was a cobbler, but his real interest lay in mathematics and

1. *George Boole: his Life and Work*, by Desmond MacHale. Boole Press, Dublin, 1985.
2. And the other? Lord Kelvin, it seems - see chapter 3.
3. 'Home-side of a scientific mind', by Mary Everest Boole. *The University Magazine*, 1878, in four parts, pp. 105, 173, 326, 454.

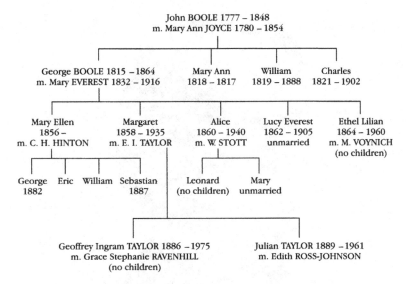

Figure 2.1 *The Boole family tree.*

in making optical instruments. His parents must, I think, have been fairly prosperous, for I have inherited some silver tablespoons dated 1789 bearing John Boole's monogram. Perhaps they were confirmation gifts, for he was only 12 in 1789. Another thing I have inherited is a box made by John Boole for a microscope that he had made. In the lid is pasted a note in my grandmother's handwriting giving this information and adding, 'He seems to have been able to do anything well except his own business of managing the shop.' The decline in John Boole's business, owing perhaps to his many outside activities, had an important effect on his son's life, for when it became clear that George had outstanding mathematical ability he was urged to go to Cambridge, but he would not do so because he wished to stay and help his father in his financial difficulties. It seems likely that the fact that Boole had no university training but absorbed his knowledge by reading great masters like Lagrange and Newton may have helped to develop the profound originality which appeared later.

John Boole was evidently a well-known character in Lincoln. He was largely instrumental in founding the Mechanics' Institute there, to give people an interest in their leisure hours, and it was partly due to his agitation that early closing of shops was

Figure 2.2 *George Boole, FRS, 1815–1864.*

enforced so that leisure was available. George helped him in this
work and when early closing was introduced, he gave a lecture
on the right use of leisure, which, though full of admirable
advice beautifully expressed, must have seemed a counsel of
perfection to his audience.

Among the optical instruments that John Boole made was a

telescope. When this was finished he put a notice in his shop, 'Anyone who wishes to observe the works of God in a spirit of reverence is invited to come in and look through my telescope'. My grandmother told me that when someone asked George's mother whether she was not proud of her famous son she said, 'Ah, but did you know his father? He *was* a philosopher.'

George had only a very simple education. After leaving his local elementary school he went for a short time to a commercial school, but his real ambition was to become an educated man. To this end he took lessons in Latin from a Lincoln bookseller, Mr. William Brooke, who became one of his closest friends. Boole then taught himself Greek and later French and German out of borrowed books. When he was 14 his father sent a verse translation of the Greek poet Meleager's 'Ode to Spring' to the local paper with a note that it was written by a boy of 14. This was printed, and was, in fact, Boole's first published work. Through the medium of the Royal Irish Academy the Lincolnshire Archives Committee has unearthed the translation from the Lincoln records and sent it to me. I showed it to Professor Donald Robertson, late professor of Greek at Cambridge, and he writes that it is a good translation in the style of Sir William Jones' translations of Persian poets in the later eighteenth century. This translation had some effect on Boole's career, for someone wrote to the editor saying that no boy of 14 could have written it. This, I think, called the attention of the people of Lincoln to the fact that they had a genius among them.

When he started to study the classics Boole wanted to be a clergyman. According to E.T. Bell who wrote an imaginative but very readable account of Boole in his *Men of Mathematics*[4], this desire was inspired by a snobbish wish to raise himself into a higher class of society than that into which he had been born. It seems to me that there is not the slightest evidence in Boole's papers or letters or in anything that his contemporaries wrote about him that that is true.[5] On the contrary, all his published non-mathematical work, as well as the accounts of those who knew him, show him to be a person whose thoughts were continually directed to, and his acts directed by, his religious

---

4. Penguin, 1937, chapter 23.
5. MacHale (*loc. cit.*) is also critical of Bell's tendency to ascribe to Boole motives for which there appears to be no evidence whatsoever.

convictions, so that it was very natural that he should want to enter the Church. It would be quite natural for a poor but clever boy to wish to use his talents to raise himself in the social scale, but I can find no evidence that this was in fact the driving motive in Boole's terrific efforts to educate himself. Contemporaries who have recorded their impressions of Boole's sense of social distinctions agree that he had none at all. My aunt (Mrs. Voynich, Boole's youngest daughter, who lives in New York) wrote me that an old lady who had known Boole in Cork in her youth told her of the following incident. 'One day in June, 1856, she went into the slum alley behind the College to engage a chimney sweep for her flues. As she was walking down the alley, she saw father ahead of her, knocking at one door after another. She came past him in time to see him passionately shaking hands with a ragged and barefoot man, and saying "I had to come and tell you, dear friends: I've got a baby and she *is* such a beauty." '

While on this subject I must comment on Bell's method of writing history. After explaining his idea that the class into which Boole was born was held in contempt by people higher in the social scale, he records his opinion of the effect on Boole of this state of affairs in the following words: 'To say that Boole's early struggles to educate himself into a station above that "it had pleased God to call him" were a fair imitation of purgatory is putting it mildly. By an act of divine providence Boole's great spirit had been assigned to the meanest class; let it stay there and stew in its own ambitious juice. Americans may like to recall that Abraham Lincoln, only six years older than Boole, had his struggle about the same time. Lincoln was not sneered at but encouraged.'

It is of course, true that Boole had a great struggle to educate himself. After a very little teaching of elementary Latin by Mr Brooke, Boole continued his lonely education by reading borrowed books, while he was teaching all day in his school. This must have been a really hard struggle. On the other hand, I know of no evidence that Boole suffered through, or indeed ever experienced, the contempt which Bell describes. There is plenty of evidence that, from an early age, Boole was encouraged and helped, to the best of their ability, by his neighbours and friends and the clergy of Lincoln. When Boole was only 19 and had been studying mathematics for two years he was asked by his fellow-townsmen to give an address about Newton on an occasion when a bust of Newton was presented to a local Institute. The address was printed at their request,[6] a fact which hardly suggests that Boole was hampered by the contempt of his fellows or those richer than himself.

Another misleading remark of the same kind occurs in Bell's account of Boole's marriage ' ... still subconsciously striving for the social respectability that he once thought Greek would confer, he married Mary Everest, niece of the Professor of Greek in Queen's College'. The word 'subconsciously' is well adapted to convey a false suggestion, for no one can say whether a subconscious thought exists or not. All that is certain is that Boole was very devoted to my grandmother – as she was to him. Other comments on this aspect of Bell's otherwise excellent account of Boole have been made by William Kneale.[7]

Boole started to read mathematics when he was 17 and was an usher at a small school. According to Mrs Boole, her husband told her that he had started to read mathematical books because they were cheaper than classical ones. My own impression is that Boole had got what he wanted out of the classics and had formed his style of English prose on classical lines. For many years after he had become interested in mathematics he used to compose poetry which was classical in form, metaphysical and religious in feeling. Two of these poems[8] have been published in accounts of his life.

Boole evidently liked writing in verse even when he had no particular urge to express poetic or metaphysical thoughts. Among his papers was found a copy in his own hand of a letter in verse which he sent to his friend Mr Brooke in 1845. It describes a visit he paid to the meeting of the British Association at Cambridge and to London and a holiday in the Isle of Wight. It is too long to quote in detail here, but I may recite a few lines to illustrate his joy in transferring very ordinary thoughts into metrical form – or as some might say, doggerel. Here are a few lines:-

6. An address on the genius and discoveries of Sir Isaac Newton, by George Boole. Lincoln, printed at the Gazette Office, 1835.
7. 'Boole and the revival of logic', by William Kneale. *Mind*, **57**, 1948, p. 149.
8. 'Sonnet to the Number Three' and 'The Fellowship of the Dead'. See Mary Everest Boole, *loc. cit.*, pp. 174 and 335.

'I made short stay in London, for I thought
With loss of rest its pleasures dearly bought
And longed in quiet indolence to be
On some lone shore of the resounding sea.'

I have a number of Boole's sonnets, some of them good, most of them not very good, but all of them sensitive to the moral and spiritual aspects of the things that inspired him.

Boole went on employing some of his leisure in writing poems after he married, but one day his wife saw him writing and asked him what he was doing. When she saw that he was writing verse she took the pages away and threw them in the fire telling him that it was bad for him to employ his time that way. He wrote no more verse after that incident, believing that the decision in this matter should rest with his wife.

Boole had a very great sense of the dignity of labour. He thought that a person who was doing a job should not be interfered with. He regarded his wife's prohibition on the writing of poetry as part of her job of making conditions favourable for his work and he never questioned it. When one of his little daughters was laying the dinner table, for instance, he would never criticize anything she did and would obey her if she told him to remove any writing or other untidy mess from the table.

Among Boole's greatest enjoyments was walking in the country near Cork and meeting the country people. In one of his vacations, spent at his home in Lincoln, Boole gave a lecture to his friends in which he drew a picture of life in an Irish cabin. This was reported verbatim in the *Lincolnshire Chronicle* at the time and I found some extracts among my grandmother's papers. It seems that Boole had taken refuge from a rainstorm in a small house beside a peat bog. He describes the house, its inhabitants, its simple furniture and the kindly welcome from the mother of the household. He notices that all the inhabitants are bare-footed. He takes off his own shoes and stockings and dries them at the fire. He continues in the style to which one becomes accustomed in reading his work: 'This denuding of the feet seems to establish a closer fellowship and sympathy; the children who have hung back come and join your circle, the dog follows them, and by and by the little pig walks up and thrusts his nose between your legs towards the flame (not however without a rebuke from the mistress) and at length the hens and domestic fowls make the social party complete.' I fear he might now have difficulty in finding these idyllic scenes.

In the early stages of Boole's study of mathematics it seems probable that he was helped by his father and by his friend, the Rev. G. S. Dickson, who had been interested in the subject since taking his degree in mathematics in Oxford. It seems, however, that Boole must have left them far behind in a very short time, for his essay on 'The Genius and Discoveries of Newton', written when he was only 19, shows that he had read and fully mastered the *Principia*. It is also clear that he had at that time read Lagrange's *Mécanique Analytique*, for he compares Newton's methods in celestial mechanics with those of Lagrange. 'By the labours of Lagrange the motions of a disturbed planet are reduced with all their complication and variety to a purely mathematical question. It then ceases to be a physical problem; the disturbed and disturbing planet are alike vanished; the ideas of time and force are at an end; the very elements of the orbit have disappeared, or only exist as arbitrary characters in a mathematical formula.' He goes on, 'In Newton's investigation this felicitous transformation could not take place. Nature must be combated on her own grounds: the disturbing force is analysed: its effect must be considered in every variety of position – above, below and in coincidence with the ecliptic plane: from syzigy to quadrature, and thence again to syzigy, the same influence is to be followed, and the resulting effects determined. The everlasting wheels of the universe are before us and their revolutions can be traced through all the changing varieties of cause, circumstance and effect.'

I think that this very early appreciation of the advantage gained by separating symbols from the physical facts they originally represented is of interest in the study of Boole's development.

Boole's first contacts with the scientific world outside his native county began when he sent a paper to the newly founded *Cambridge Mathematical Journal* in 1839. In the first volume of this journal the names of authors are not printed, each article being signed by a letter or symbol which told the initiated who the author was. The convention was probably adopted because in Cambridge at that time it was thought that to do scientific work was a fitting amusement for a gentleman but that to publish it savoured of self-advertisement, indeed the late Dr J.W.L. Glaisher told me that this idea was still prevalent when he was young in the 1860's. In the second volume of the *Cambridge Journal* this convention was dropped and Boole's first scientific paper on analytical transformation appeared in it under his name in 1840. This paper and one published in the following year on the theory of linear transformations seem to have been precursors of the theory of invariants which, in the hands of Cayley and Sylvester, became so important later. During the same year Boole published his paper on the 'Integration of linear differential equations with constant coefficients'. Though the use of the symbol d/dx as though it were an algebraic number had been touched on by Herschel and others it was this paper of Boole's which first brought out the great power of symbolic methods in mathematics.

Between 1840 and 1842 Boole rapidly developed his ideas on this subject and in 1842 he presented his 'General method of analysis' to the Royal Society. It was for this paper, when he was only 28, that the Society awarded him one of the Royal Medals in 1844. In it he shows the immense power of this method and applies it to a great variety of mathematical subjects. He concludes his paper with the paragraph:

'The position which I am most anxious to establish is that any great advance in the higher analysis must be sought for by an increased attention to the laws of the combination of symbols. The value of this principle can scarcely be overrated and I only regret that in the absence of books and under circumstances unfavourable for mathematical investigation I have not been able to do that justice to it which its importance deserves.'

At that time (1844) the gold Royal Medal was accompanied by a silver replica. I was left this silver replica by my grandmother who had had to sell the gold one. I have compared it with the Royal Medal given to me 89 years later and find that the design is unchanged and the modelling is still perfect.

In 1849 Boole was advised by his Cambridge friends to apply for the professorship of mathematics at the newly-founded Queen's College, Cork. With great perspicacity the authorities appointed him though he had no University training or degree.

Boole took a great deal of trouble to make his lectures clear and he published text-books on differential equations and on finite differences which give an idea of his teaching methods. Ultimately his conscientiousness in this respect led to his death for he got wet on his way to a lecture and insisted on lecturing in soaking clothes, an act which gave rise to his last fatal illness in 1864. His pupils were devoted to him. One day he arrived in his lecture room before them and started walking up and down in front of his blackboard thinking out some problem. The students arrived and sat down waiting for him to begin, not liking to interrupt his thoughts. After an hour they left, Boole still walking in front of the blackboard. When he got home he said to my grandmother, 'My dear, a most extraordinary thing happened today. None of my students came to my lecture.'

Mary Boole

Mary Everest, born in 1832, was the daughter of Thomas Everest, a clergyman who lived in Gloucestershire. Her father's brother, Sir George Everest, after whom the mountain was named, was a soldier serving in India who became the Surveyor-General and one of the founders of geodesy. George Everest was attracted by the eager and strong character of his young niece Mary, and in later life she recorded that he wanted to adopt her. However, she did not wish to be separated from her parents. Later she met George Boole while visiting Cork and staying with her mother's brother, Dr Ryall, Vice-President and Professor of Greek at Queen's College (who had supported strongly Boole's candidature as Professor of Mathematics on the recommendation of a friend at Cambridge). Mary Everest and George Boole became friends, and he introduced her to mathematics. After the death of her father in 1855 they were married. Their married life in Cork lasted for the very short period of nine years, for in December 1864 Boole died of a chest ailment leaving his widow Mary, aged 32, with five very young daughters, the oldest being only eight. Figure 2.3 shows Mary Boole's family some years later.

Mary Boole was evidently a very able woman and in particular an accomplished writer, although she held some rather eccentric ideas on education. After their marriage George continued to teach her mathematics, and she used this knowledge to relieve him of many of the more straightforward parts of his mathematical work such as proof correction and the working out of illustrative examples and the clarification of mathematical arguments. More generally she regarded her role as being to protect him from influences which might hinder or diminish his creativity. The long and discursive article in four parts entitled 'Home-side of a scientific mind'[9] that she published some years after George's death gives an interesting picture of an unworldly scholar in the care of a protective woman who managed

9. Mary Everest Boole, *loc.cit.*

her husband, house and children and who found time for discussion of the religious issues of the day with which George seemed to be greatly concerned.

At the end of the article on George Boole by Geoffrey Taylor which has already been reproduced here in part (1954f), Taylor wrote as follows:

> I should not end this note without saying a few words about my grandmother. She must always have been a person who saw only one thing at a time, very intensely and very large. During her married life and for some time after it, I think her husband filled

Figure 2.3 *George Boole's wife Mary, née Everest, and her five daughters (from left to right, Margaret Taylor, Ethel Lilian Voynich, Alice Stott, Lucy Boole, Mary Hinton) and five of her grandchildren (Julian and Geoffrey Taylor, Mary and Leonard Stott, George Hinton). c. 1894.*

her whole world. As I have mentioned, she wrote a disjointed but very revealing account of her life with Boole and though her devotion to him is obvious, it did not blind her to what she regarded as his faults and peculiarities. She expressed the opinion that while the more educated people in Cork regarded her husband as some kind of saint, the poorer people thought of him as so simple that it would be unsporting to try and cheat him. She writes that an old lady once said to her, 'They tell me your husband is a very learned man, my dear. I never saw a learned man before that one could talk to. He's as innocent as a child, bless him, and as good as an angel, but I never should have guessed that he was clever.'

In later life Mrs Boole wrote on the education of children and I think that many of her ideas have been adopted, but unfortunately she began to believe herself to be an interpreter of Boole's work which she regarded as having a psychological meaning.[10] She had a large number of friends who filled the role of disciples and one of them eventually collected her writings[11] which were published in four volumes. I had, I fear, little interest in the things she wrote about but I visited her very frequently and enjoyed her conversation, particularly when she described her childhood in a Gloucestershire vicarage and in Paris.

While George Boole was alive he and his family lived comfortably in a fine house in Cork with at least two servants, but his death on 8 December 1864 put the family in a difficult financial position. Mary Boole's roots were in England, and fortunately an opening for her at Queen's College in Harley Street, London, materialized. George had been a friend and admirer of F.D. Maurice, a clergyman in London who was one of the founders of this early women's college, and it was presumably through his good offices that in 1865 Mary was offered the post of librarian at Queen's College and the tenancy of one of the College houses. Mary later held a variety of positions in

10. See *Mary Everest Boole – A Memoir with Some Letters*, by Eleanor Cobham. Ashingdon, Essex, 1951.
11. Including the following articles: The Message of Psychic Science to Mothers and Nurses, 1883; Symbolical Methods of Study, 1884; The Mathematical Psychology of Gatry and Boole, 1897; Logic Taught by Love, 1905; Mistletoe and Olive, 1908; The Forging of Passion into Power, 1912.

the College, including teaching mathematics, and all her daughters except the youngest attended the school attached to the College. She also did freelance teaching, to judge by her advertisement in the first issue of *Nature*, dated 4 November 1869, soliciting female 'pupils in arithmetic, algebra, analytical geometry, the differential calculus, etc.' The minutes of the meetings of the Council of Queen's College record that she took leave of absence from her College duties from mid-March 1874 owing to illness, and in the following year when, according to the Council minutes, 'her illness had assumed the form of temporary derangement', she resigned from her positions.

The five Boole daughters

The five daughters of George and Mary Boole all became women of character, and three of them – Alice, Lucy and Ethel Lillian – exhibited unusual ability in science or art. The second oldest daughter, Margaret, was the mother of Geoffrey Taylor.

Boole's third daughter, Alice, who was only four when he died, became an amateur mathematician like her father. Taylor described her work on four-dimensional geometry in two articles of reminiscences late in his life (1963c, 1969d). Here is an extract from the latter of the two, entitled 'Amateur Scientists' (among whom Taylor included himself, of which more later), which was delivered as an address at the Sesquicentennial Celebration of the University of Michigan:

> Alice had no mathematical training beyond the books of Euclid, but her oldest sister's husband, Charles Howard Hinton, an imaginative teacher, though not a very competent mathematician, interested her in four-dimensional geometry, which he thought about in much the same way as did Edwin A. Abbott in his *Flatland*. In that romance Abbott described the perceptions of flat beings who could not move out of a plane. Hinton asked himself 'How can we develop a capacity for appreciating four dimensions with our senses?' But Alice Boole asked herself the much more definite question, 'What shape would we see if a four-dimensional body passes through our three-dimensional space?' She was particularly interested in the convex regular figures, of which there are five in three dimensions: the tetrahedron, cube, octahedron, dodecahedron,

and icosahedron. These figures are bounded by regular two-dimensional polygons, the triangle, the square, and the pentagon. In four dimensions, an analogous regular figure, or polytope as it is called, must be bounded by a number of regular three-dimensional figures. Alice found that there are six of them, and they are bounded by 5, 16 or 600 tetrahedra, 8 cubes, 24 octahedra or 120 dodecahedra.

I have no doubt this was known by analysts before, but Alice's method of discovery was typically that of an amateur. She started by noticing that a corner in a regular four-dimensional figure bounded by tetrahedra, for instance, can only have either 4, 8 or 20 of them meeting at a point because a section of three-dimensional space close to the corner in a symmetrical position could only be a tetrahedron, an octahedron, or an icosahedron. She then traced, using only Euclid's construction, the progress of the section as the four-dimensional figure passed through our three-dimensional space.

In this way Alice, employing only Euclid's constructions, produced sections of all the six regular polytopes. She found, for instance, that the figure for which 20 tetrahedra meet at a point is

Figure 2.4 *Alice Boole's cardboard polytopal sections.*

bounded by 600 tetrahedra. She constructed sets of sections in cardboard of this figure counting the tetrahedra as they passed through our space till the section closed after 600 had passed. After completing this really astonishing feat, the idea of publishing any of her results never entered her head. She had married and completely gave up her mathematical hobby to look after her children. About ten years later, one of her friends noticed in the *Proceedings of the Amsterdam Academy* a paper by Professor Schoute of Groningen in which he calculated, by purely analytical methods, the central sections of the regular polytopes. Schoute's projections of these were identical with Alice's so she erected complete sets of her cardboard models, photographed them and sent prints to him. The photograph [in figure 2.4] is one of these. It shows sections of the 600 cell and the polytope bounded by 120 dodecahedra. Schoute could not understand how Alice had found the non-central sections any more than she could understand his analytical methods. But they became great friends and used to visit together Ethel Everest, a daughter of Sir George, in order to collaborate in further work. Schoute insisted on Alice publishing her results, which she did under her married name, Alice Stott.

Boole's fourth daughter, Lucy, studied chemistry and was appointed to a lectureship at the London School of Medicine for Women. She lived with her mother in London and died at the age of 43, unmarried. Taylor records that in her short life she collaborated and published with Wyndham Dunstan and, so he believed, became the first woman professor of chemistry in England, at the Royal Free Hospital Medical School in London.

The youngest daughter, Ethel Lillian, was a high-spirited girl who did not get on well with her mother. She seems to have led a life quite different from that of her sisters. At an early age she went to Berlin to study music for three years and on her return became involved in a London-based group of Russian revolutionary workers. She learnt Russian, and in 1887 went to St Petersburg and other places in Russia for two years in order to help the families of political prisoners. On the outward journey she passed through Warsaw, and went to the Warsaw Citadel, where some of the revolutionaries whose families she proposed to help were being held as prisoners. Later, in 1892 in London, she met Wilfred Voynich, a Polish revolutionary who had just escaped from exile in Siberia and who is said to have declared that he remembered looking down on the prison yard and seeing this golden-haired young woman in a black cloak from the window of his cell on Easter Sunday in 1887. A year later they were married.

Ethel Lillian made use of all these experiences in her first novel, *The Gadfly*, published in 1897, which, unknown to the author, became phenomenally popular in the communist countries, especially in the former Soviet Union, where several million copies were sold and the author was ranked with Dickens. It is a powerful and tragic story of two young lovers, she English and he Italian, engaged in revolutionary activities in Italy in the mid-nineteenth century, and of the rejection of religion by one of them. This tale is made even more romantic by the claim made in Bruce Lockhart's book *The Ace of Spies* (Hodder & Stoughton 1967) that the model for the young Italian man was his father's friend, Sidney Reilly, later to become Britain's master spy, with whom Ethel Lillian herself had a brief affair during the summer of 1895 in Italy. The extraordinary story of Ethel Lillian's life as a young woman and her first novel, an intimate mixture of fact and fiction, is told by Anne Fremantle in an article entitled 'The Russian best-seller'.[12]

Wilfred Voynich later became a rare-book dealer, and the Voynichs moved to New York where Ethel Lillian lived from 1920 until her death in 1960. She published several other novels, translations of Russian and Ukrainian literature, and some music, although nothing comparable in success with that of *The Gadfly*. Taylor visited his Aunt Ethel Lillian in New York several times in the 1930s and the 1950s, and always found her very lively and interesting.

The paternal lineage

A few generations of Taylor's family tree on his father's side are known. There seems not to be any evidence of outstanding intellectual ability before the tree reaches Geoffrey Taylor, but both his father,

---

12. *History Today*, Sept. 1975, pp. 629-37.

Edward Ingram Taylor, and his grandfather, James Taylor, were interesting characters.

James Taylor was a foundling and was regarded within the family as a mysterious figure. Geoffrey Ingram Taylor wrote in his Royal Society record that nothing is known about Edward Ingram Taylor's ancestry, except that his father was a foundling. That is no longer true, because there was a document issued by the Foundling Hospital in London (full title: Hospital for the Maintenance and Education of Exposed and Deserted Young Children) among the papers left by Geoffrey Taylor on his death in 1975 which led to quite a lot of information. My wife Wilma knows her way around the social welfare field, and she found by enquiry that the modern successor to the Foundling Hospital is the Thomas Coram Foundation for children at Brunswick Square, London, and that records of children in the care of the Foundling Hospital in the early nineteenth century are held at the Greater London Record Office in the old County Hall. With the help of an archivist there, Wilma compiled the following account of the life of James Taylor:

> James Taylor, born 20th September 1807, was the illegitimate son of Sarah Powell and Richard Ort. Sarah was a servant in the house of Mrs Benedict, no. 4 Great Mays Building, St Martins Lane, London; Richard Ort was a fellow servant and a journeyman to Mr Benedict who was a bookbinder. Richard promised marriage to Sarah, but disappeared when she was three months pregnant and was not heard of again. Sarah 'laid in' with Mrs Wood, Horse & Groom yard nr. Leather Lane, Holborn. She petitioned the governors of the Foundling Hospital on 24th November to take her baby son who was then between two and three months old. She had pawned her clothes and did not know 'how to procure subsistence for herself and child through illness and distress of mind'. The child was admitted to the Foundling Hospital on 19th December, 1807, as no. 18855. A grant was also given to Sarah by the governors for clothes. Mrs Benedict reported her as an honest, good and sober servant and was willing to take her back. Sarah's petitions were signed with her mark. (See figure 2.5.)
> Foundlings were given names by the hospital, those of governors and benefactors often being used. There is evidence of

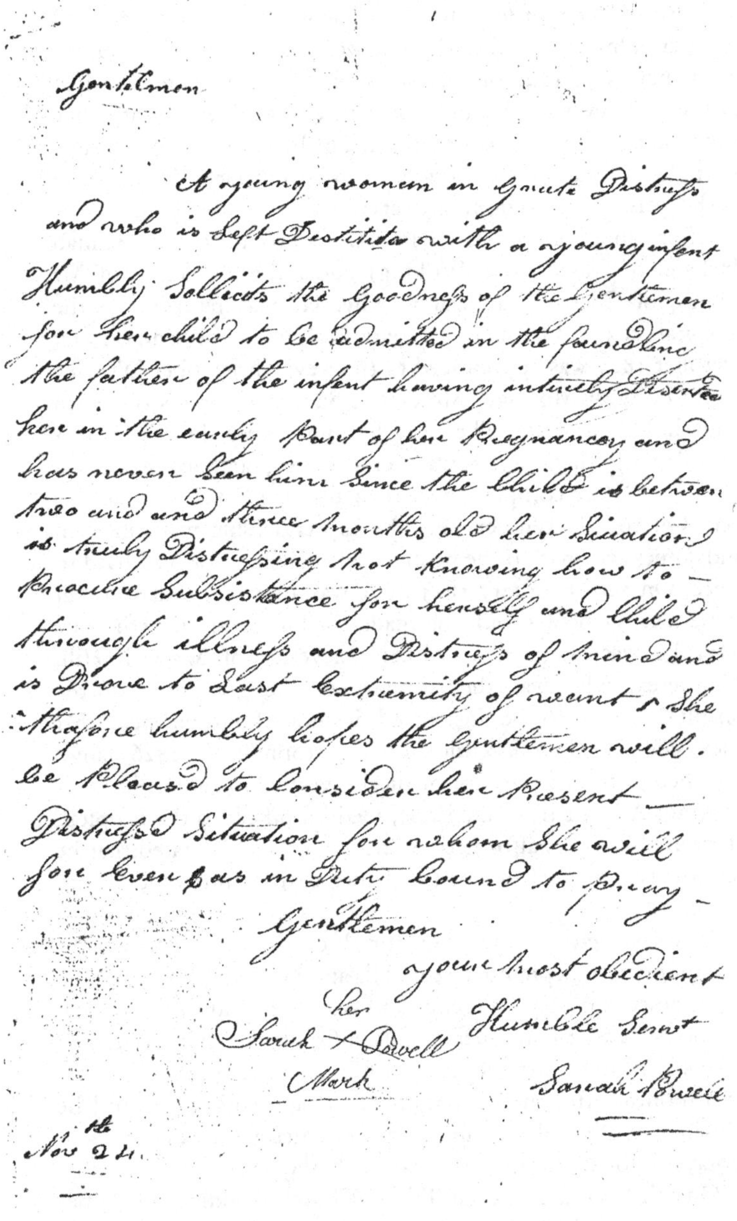

Gentlemen

A young woman in Great Distress
and who is Left Destitute with a young infant
Humbly Sollicits the Goodness of the Gentlemen
for her child to be admitted in the foundling
the father of the infant having intirely Deserted
her in the early Part of her Pregnancey and
has never seen him Since the Child is between
two and and three months old her Situation
is truly Distressing not Knowing how to —
Procure Subsistance for herself and Child
through illness and Distress of mind and
is Drove to Last Extremity of want r She
therefore humbly hopes the Gentlemen will.
be Pleasd to Consider her Present —
Distressd Situation for whom She will
for Ever pas in Duty bound to Pray —

Gentlemen

your most obedient

her
Sarah + Powell   Humble Servt
Mark   Sarah Powell

Novr 24.

Figure 2.5 *Sarah Powell's petition to the governors of the Foundling Hospital on 24 November 1807.*

23

a Sir Henry Taylor having a connection with the hospital, but no suggestion either way that his name was used in this case. All children admitted to the hospital were fostered in the country until they were four or five years old, when they returned to London. They were brought up and educated there until they were old enough to be apprenticed. The boys went into a variety of trades although many became Army bandsmen. The majority of the girls went into domestic service.

After admission, James Taylor was sent to nurse with Hannah Jackson of Oaking (i.e. Woking) under the inspection of Mr Living of Chertsey, a local worthy. He was returned to the hospital on 30th October 1813 (later than was usual) and on 31st October 1823 was apprenticed to the Rev. Christopher D'Oyley Aplin of West Moulsey (Molesey) , Surrey, to be instructed in household business. The Rev. Aplin wrote in November complaining that James was scrophulous, and again on 1st December 1823 complaining that James continually fouled his bed. On 10th December 1823 the apprenticeship was cancelled and James returned to the hospital. He may have proved hard to place. On 16th February 1825 there was a recommendation that some arrangement should be made for him, and on 15th June 1825 he was assigned to Morris Lievesley of Muswell Hill, Middlesex, who was the secretary to the governors of the hospital, again to be instructed in household business. Mr Lievesley also drew attention to his scrophula. In 1826 James returned to the hospital for medical investigation about his bed-wetting. A letter from Dr Earle, medical officer to the hospital, attributed it to a malformation of the urethra and stated that he had devised an instrument for James to use and could do no more.

James served Mr Lievesley for three years and six months ending on 20th September 1828. His apprenticeship was 'marked by honesty, sobriety, diligence and attention, and to the best of his power under continued bodily infirmity he endeavoured to give satisfaction'. Mr Lievesley continued him in his service for two months after the expiration of his apprenticeship until he obtained another place. James was presented with a gratuity and a prayer book by the governors on 31st May 1829.

On 8th February 1853 at Trinity Church, Paddington, James Taylor married Matilda Ingram of 20 Weymouth Street, daughter of William Ingram, silversmith, and Elizabeth née Parrott. His occupation was given as servant, and his address as 55 Westbourne Terrace. Two children were born of the marriage, Harriett Elizabeth (born 11th July 1854, died 10th May 1883 of opium poisoning, with a coroner's verdict of misadventure) and Edward Ingram (born 10th November 1855, died 2nd August 1923) who married Margaret Boole on 22nd April 1885 and was the father of Geoffrey Ingram Taylor. James became a lodging-house keeper at 3 Radnor Place, Paddington. He died on 14th October 1874, and as he had purchased a burial site in 1873 at Paddington Cemetery presumably he was buried there, in grave 3957. His wife Matilda died on 1st December 1872.

Photographs of James and Matilda Taylor are reproduced in figure 2.6. There exists also a fine painting in oils of Matilda Ingram, who was evidently a good-looking woman, made in the year in which she married James Taylor.

Edward Ingram Taylor (born 1855) showed artistic skill at an early age, and studied at the Slade School of Art after leaving school. At the West London School of Art he was awarded a travelling studentship which enabled him to study decoration and design abroad for two months. This coincided with his marriage in 1885 to Margaret Boole, who was one of his art pupils, and their visit to Italy was partly a honeymoon and partly an art course. Figure 2.7 shows the married couple in later years.

Throughout his married life Edward Taylor supported his family mainly by undertaking commissions for the design of stained glass and the design of the decoration of large public rooms. He developed a connection with Harland and Wolff, ship-builders at Belfast, and had many commissions to plan the decoration of the public rooms on passenger liners and to paint large pictures for these rooms. His chief love was landscape painting, and he gave all his free time to this. Some of his landscapes were exhibited at the Royal Academy. Later in life he spent most of his leisure time making small pencil drawings of English flowers (see figure 2.8). These are exquisite, and examples may be seen in the British Museum and the Natural History Museum in London, and the Fitzwilliam Museum, Cambridge.

Geoffrey Taylor recorded that although his father worked very hard all his life, he took a great many holidays walking and painting in the country. He was devoted in particular to the Thames for rowing

(a) (b)

Figure 2.6 *(a) James Taylor 1807–1874, grandfather of Geoffrey; (b) Matilda, née Ingram, 1816–1872, m. James Taylor.*

and painting. He usually took his family on these holidays, and often school friends of his two sons. In later years Geoffrey reciprocated by introducing his father to the pleasures of sailing; in 1922 they sailed across the North Sea alone in Geoffrey's boat *Sorata*. Edward Taylor died one year later.

The family house at St John's Wood, London, was both studio for Edward Taylor's work, and home for himself and his wife Margaret and the two boys. Many other artists lived and worked in the neighbourhood. They were a convivial lot, and Geoffrey Taylor described his father's artist friends as the most cheerful people he ever met. It was a friendly, happy and interesting family environment.

a) 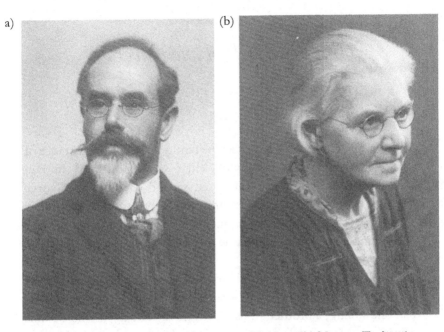 (b)

Figure 2.7 *(a) Edward Ingram Taylor, father of Geoffrey; (b) Margaret Taylor, née Boole, mother of Geoffrey.*

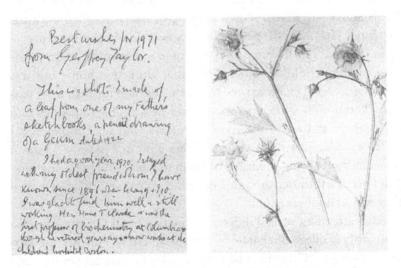

Figure 2.8 *A Christmas card for friends improvised by G.I. and incorporating one of his father's delicate sketches of wild flowers. (Reproduced by permission of one of the recipients, Milton Van Dyke.)*

CHAPTER 3

Childhood, school and university

The preceding chapter sets the family stage for the entrance of Geoffrey Ingram Taylor, elder son of Edward Ingram Taylor, artist, and Margaret, second daughter of George Boole. It is already an illustrious family, with several notable members showing a clear disposition to independence and originality. The new member soon showed that he too was well endowed with those characteristics.

Geoffrey, as we may now call him, was born on 7 March 1886 at 10 Blenheim Villas, St John's Wood, an inner suburb of London. He had one sibling, a brother Julian, born three years later, who became a surgeon. Mary Boole and some of her daughters lived in London during Geoffrey's childhood and youth, and they and his father's artist friends all contributed to the family influences. Alice Stott, Lucy Boole and Ethel Lillian Voynich in particular were often with the two boys. However, the parental love was the dominant influence.

In later life Geoffrey wrote: 'My home meant much more to me than anything I did at school. My father worked at home and was able to devote much more time to my brother and me than most parents. Though his work was poorly paid he was able to take us into the country and on to the river frequently, and my brother and I were never happier than when we were with our parents. I cannot remember ever being punished by them or being told to do anything I did not want to do, though I do remember doing things I hoped they would not hear about'.

After attending several different primary schools Geoffrey went to University College Preparatory School at Holly Hill for the two years 1897–99. While there he made friends outside his family circle for the first time. He also began to take an interest in things outside the range of play and woodwork. In particular he acquired an interest in wild flowers, which remained with him for the rest of his life (as

Figure 3.1 *Geoffrey and his younger brother Julian with their parents at the family home in London, date unknown.*

those who visited his house 'Farmfield' in Cambridge discovered). The first stirrings of interest in mathematics and science at school also appeared ('I was fascinated by Euclid'), although Geoffrey could not later remember any individual teacher being responsible for this.

A development outside school in 1897 appears to have provided a quite decisive pointer to the direction of his later life. A friend of Geoffrey's mother, Dr Helen Webb, gave her two tickets for the children's Christmas lectures at the Royal Institution (an annual event, initiated by Faraday in 1827). The lectures at Christmas 1897 were delivered by Sir Oliver Lodge on wireless telegraphy, and Geoffrey and a school friend named Maurice Perrin attended. The lectures and demonstrations fascinated Geoffrey, and over 50 years later he described his reaction in a way which will strike a chord with many scientists: 'I wish I could again capture the exquisite thrill those lectures gave me. From that time I knew I wanted to be a scientist' (1952d). The two boys followed up the lectures with some enterprising experiments of their own. They constructed a Wimshurst electric 'influence' machine for the production of electric charge, and used it to generate X-rays in the manner demonstrated by Rontgen just two years previously. For this purpose Geoffrey's aunt Lucy Boole lent him a bulb, made by a glass-blowing technician according to Röntgen's original specification. They found that the bulb emitted a green light when connected to the two poles of the Wimshurst machine and that X-rays were generated. Geoffrey's mother obligingly kept her hand still on a frame containing a photographic plate while the two boys turned the handle of the Wimshurst machine for 45 minutes. The photograph turned out to be a good one, and revealed some arthritis.

Lord Kelvin was present at one of these Christmas lectures at the Royal Institution in 1897, and Geoffrey was introduced to him as a grandson of George Boole. Many years later (1963c, 1964c) Geoffrey wrote as follows: 'I have a faint memory of a kindly and bearded old gentleman looking down at me and telling me something about the laying of the first Atlantic cable, which I was too shy to admit I had not understood. I also remember him saying that Boole had been one of the people from whom he had asked a recommendation when he applied for his professorship at Glasgow in 1846. I had always suspected that I must have got confused or misheard Kelvin, because Boole was only 30 at the time and has always been known as a very pure mathematician, the principal founder of symbolic methods in mathematics and logic and of

Boolean algebra. One would hardly expect the young Mr William Thomson, as Kelvin then was, to ask for a recommendation for the professorship of natural philosophy from such a person, particularly when he was only a young schoolmaster in a small elementary school. However, some years ago I had to say something about Kelvin at a meeting of the Institution of Civil Engineering[1] in London, and I thought it just worthwhile to search Sylvanus P. Thompson's *Life of Lord Kelvin* for a possible reference to his sponsors in his application for the Glasgow Professorship. Sure enough, the recommendation of G. Boole Esquire was quoted, and I was particularly impressed to learn that the paper of Kelvin's which Boole picked out as having interested him was one on hydrodynamics, a subject which has always fascinated me'.

Geoffrey also found that the Department of Natural Philosophy at Glasgow University has a collection of 15 letters from Kelvin to Boole written in 1845–48, from which it appears that they were friends with a common interest in mathematics, that Kelvin wanted to attract contributions from Boole for the new *Mathematical Journal*, and that Kelvin reciprocated by recommending Boole for some positions in Ireland.

In 1899 Geoffrey won a scholarship which took him to University College School, much to the astonishment of his teacher at the UC Prep. School, who knew only that Geoffrey was bad at languages. University College School was then in Gower Street, and the Headmaster, J.L. Paton, whom Geoffrey later described as 'inspiring', was a classicist whose real interest was in educating boys. The admiration was evidently mutual, to judge by a perceptive testimonial written by Paton after he left UC School to become High-master of Manchester Grammar School in 1905: 'Geoffrey Ingram Taylor was under me for nearly 5 years at University College School. He was a hard worker and keen on his studies. He was also very successful. But what struck me most about him was his ability to work a subject out by himself, in practical work as well as in book-work. He has the student mind and I have had no boy of whom I

1. From which Geoffrey Taylor received the Kelvin Medal for 1959.

would say with more confidence that he was likely to excel in original research'.

When University College School was founded, about 1830, it was the only school in England which exercised no religious discrimination, with the consequence that there was a high proportion of Jews and Unitarians among the pupils. The academic standards were high, and for some years only Eton had produced more 'senior wranglers' (boys who got the most marks in the annual examination in mathematics at Cambridge). The senior boys were often allowed to attend lectures given at University College, and Geoffrey found fascinating a course given by Sir Arthur Evans on his recent discovery of Knossos. Another facility which Geoffrey appreciated was a practical course in the workshop of the Engineering Department. When he went there some 30 years later with his friend G.T.R. Hill, who had just been appointed to a professorship at UCL, they found that the little steam engine which the two of them had made in the workshop was still being used in the teaching programme.

A significant incident in Geoffrey's boyhood was later recorded in the following words: 'When the relief of Mafeking was announced in May 1900 the school was given a holiday, and most of the boys walked down to Hyde Park Corner and cheered outside Baden Powell's house. I think it was then that I first began to think that the country is over-populated. There were very large crowds and I felt uncomfortable in them. I have always disliked being jostled in a crowd, and perhaps that was why I disliked football so much'.

Geoffrey's interest in sailing in small boats, later to become a passion, developed during his youth and led to an ambitious enterprise to build his own boat during his last year at school (1904–5). He included the following account of his first taste of sailing in a review of his life as what he thought of as an 'amateur scientist' (1969d):

> Ever since as a small boy I set out to cross a pond in a tub, I have wanted to sail, though none of my family or friends were inclined that way. When I was at school, I was determined to build a boat. No ready-cut kits were then available of course, so I had to design my boat myself, one that I could make in my bedroom with ordinary materials from a house-builder's yard.

Figure 3.2 *The boat built by Geoffrey in his bedroom.*

My bedroom was 14 ft by 10 ft, on the second floor. My boat was 13 ft 6 in by 5 ft 3 in wide, and my bedroom window was about six inches narrower. People were always telling me of Robinson Crusoe's famous unmovable boat.

But I had my plan. I rigged a joist out from the window of my parent's bedroom, fixing the inner end under my parents' bed, and fixed a borrowed builder's pulley at the outer end. I took out the sash-window, put slings round the boat, turned it on its side, and made the slings fast to a rope, which I led through the pulley and down into the garden. Then my small brother, a cousin and I hung on to the rope, while a grown-up pushed the boat out of the

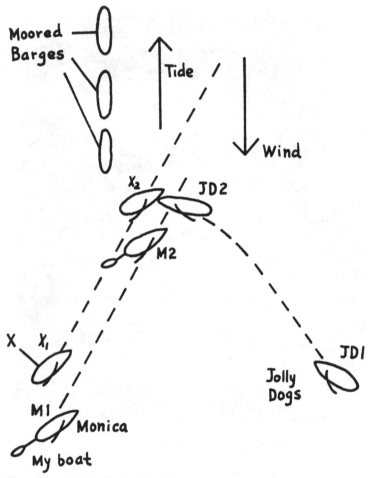

Figure 3.3 *Geoffrey Taylor's first cruise was on the Thames in 1905. His sailing dinghy is being towed here by the sailing barge Monica. X and JD are other barges. Suffix 1 indicates the pre-collision positions of the three barges, and suffix 2 their positions at collision.*

window and turned it flat against the wall of the house. We began letting out the rope. Suddenly the boat slipped a little in the slings, my brother let go to run up and fix them, and my cousin and I started to rise. Fortunately, we held on, and my brother seized us before we had gone up very far. Then we got the rope round the handle of a heavy garden roller, and the rest of the operation proceeded according to plan. (see figure 3.2.)

My mother made the sails, we launched on the Thames, and

I sailed down to the sea, sleeping at night, very uncomfortably, with one leg down each side of the centreboard case. Returning, I was making little headway against a tiresome sea and head wind. Some fifty or sixty sailing barges were beating up with the tide, and I decided to try to get a tow. When the barge comes round at the end of a windward tack, it very nearly stops altogether. That was my moment. I got near the bank where the barges had to come about, and, after two failures, got a tow rope aboard a barge called the *Monica*. After some time, we came upon a number of barges anchored mid-stream. We were sailing close-hauled on the port tack (wind on our left). Normally a boat on the port tack must give way to one on the starboard, but the rule of the road gives a close-hauled boat on the port tack the right of way over one on the starboard tack with the wind free. As we got near the anchored barges, *Jolly Dogs*, a sailing barge on the starboard tack, came up on a collision course with a barge (X) just ahead of us on the port tack. *Jolly Dogs* could have held on and forced X and *Monica* to pass under her stern, but then would have had to tack immediately after passing us to avoid the moored barges. Or she could have fallen off, thus losing her right of way, and passing astern of us. Unfortunately, she did not bear away till too late. The positions of the barges at this moment are shown as $M1$, $X1$, $JD1$ in the diagram (figure 3.3). In a short time there was a resounding crash as X and JD collided and *Monica* ran into them both. The collision positions are shown as $M2$, $X2$ and $JD2$. The verbal exchanges between the three skippers and their mates when we were interlocked, and before X ultimately sank, were very educative. I listened with admiration, and tried to remember some of the more imaginative phrases. I did not think I had, till many years later, when I was entering a lock in my forty-eight foot yacht, the man who should have snubbed my progress failed to do so, and I thought I was going to ram the lock gates. Out they came. A lady watching from the lock wall was seen to pick up her Pekinese, stop up its ears, and carry it out of earshot.[2]

2. Geoffrey Taylor wrote these words, but I cannot believe what the words say. I doubt if anyone ever heard him use 'bad' language, colourful or otherwise; and the donnish joke about the dog looks like pure invention. He was not a natural raconteur in person, although on paper his innocent gentle humour with characteristic British understatement was appealing. But why should he make up this story?

Well, *Jolly Dogs* had been clearly in the wrong, but she blamed *Monica* and a pretty case was brewing for the Law Courts. The lawyer for *Monica* asked me to describe what happened. I could give a consistent account and agreed to give evidence in court. Soon afterwards, the lawyer wrote me that the opposition had dropped their allegations when they learned that he had an independent witness, even though a schoolboy, not a bargee, who could describe the collision. The amateur sailor's knowledge had proved valuable.

Geoffrey's attraction to science and mathematics was evident during his years at University College School, and appears to have been both theoretical and practical. Later he wrote, 'While still at school I came across Lamb's *Hydrodynamics* in my Uncle Walter Stott's library, and though I could not understand it I was fascinated by its subject and hoped that I would some day be able to use it in understanding the mechanics of sailing boats, a subject I was already much interested in from the practical point of view.' In 1904 his teachers advised him to try for a scholarship at Cambridge by taking the open examination. A testimonial from a mathematics teacher at the school described him as the most impressive pupil of the previous 15 years, and another as 'the most promising pupil that the School has had for many years'. It is thus not surprising that his application to Trinity College, which he chose for its association with great names like Newton, Maxwell and J.J. Thomson, led to the award of an 'exhibition'. In later years (1952d), Geoffrey was moved to defend the system of awards: 'If you read the works of politicians and educational reformers you might believe that Cambridge 50 years ago was a preserve of the rich. This is quite untrue. Though some of the students were wealthy it was possible for almost anyone who was capable of benefiting by the education there to go to the University. My Father's painting of landscapes was not a very lucrative occupation so, though he was willing to make any sacrifice on behalf of my brother and me, he could not afford to send us to college without assistance. I therefore applied to my local County Council for funds to enable me to take up my scholarship. The Council made an estimate of what I would need, subtracted the value of my scholarship and a small sum which they estimated my parents might be expected to contribute and gave me a grant to make up the difference. This was the procedure which they adopted in my district in all cases where school-boys obtained open scholarships by examination. Of course many boys of real ability might not be able to get scholarships but this was usually due to difficulty in studying under adverse home conditions rather than a lack of educational facilities'.

In October 1905 Geoffrey went up to Trinity College and began a new life as an undergraduate. He chose to study mathematics for the first two academic years, since this seemed to him to be the gateway to the physical sciences. The mathematics lecturers in Trinity at that time were A.N. Whitehead, R.A. Herman, E.T. Whittaker, E.W. Barnes (who later became Bishop of Birmingham), and G.H. Hardy, who replaced Whittaker in 1906. Later Geoffrey wrote (1952d): 'I had never cared for pure mathematics, and though I attended Barnes' lectures I did not follow them carefully. The lectures I enjoyed most in my first year were on what must be intrinsically the dullest branch of mathematics – geometrical optics – but they were given by Whitehead who would make any subject interesting, even philosophy.'

The Mathematical Tripos in those days was a very keenly contested examination for which many of the candidates were coached personally. After two years Geoffrey took Part I of the Mathematical Tripos.[3] Candidates for this examination were put into Class 1, 2 or 3 according to the marks gained, and were also ordered within each class. Candidates put into the top class 1 were called 'Wranglers', and the one who was at the top of the list was given the coveted title 'Senior Wrangler' (until 1909). At the other extreme the one who came at the bottom of class 3 was always given a wooden spoon. The drilling and constant practice needed for a high place in the

3. I have looked at the examination papers for Part I which Geoffrey took in 1907 and find the questions formidable, in that they require sharp wits rather than knowledge. There is an emphasis on Euclidian geometry which today's teachers would find strange but which may well have suited Geoffrey's ability to visualize three-dimensional relationships (like that of his Aunt Alice). He had a gift for mechanical design (exemplified by the C.Q.R. anchor), and preferred geometrical constructions to algebraic analysis.

Figure 3.4 *Geoffrey, growing up. (a) Date unknown; (b) date unknown; (c) at age 19, in 1905; (d) at age 25, in 1911.*

mark list would not have appealed to Geoffrey, and his presence on the river rowing in all the twice-yearly boat races and sculling often for pleasure (like his father) could not have helped; in the event he was 22nd Wrangler.[4]

At the beginning of his third year Geoffrey transferred to Part II of the Natural Sciences Tripos and studied physics in the Cavendish Laboratory. He attended lectures by C.T.R. Wilson (about whom Geoffrey later said 'I soon realized I had the good fortune to be under a great man'), and J.J. Thomson whom he also found to be 'inspiring'. He obtained first-class honours in the final examination in 1908, and graduated as B.A. Most importantly he was awarded a major scholarship at Trinity College which enabled him to stay in Cambridge for research. The awarding of that scholarship launched a remarkable research career and led directly to lifelong associations with Trinity College and the Cavendish Laboratory.

4. Geoffrey was one of the last to be subjected to this published ordering, which was abolished in 1909. The term 'Wrangler' was subsequently used instead to indicate those candidates who obtained first-class honours in Part II of the Mathematical Tripos, as it is at present.

CHAPTER 4

First steps in research

There was no PhD degree at Cambridge in the early years of this century. Students who did well in their final undergraduate examination and wished to take up research training simply asked a Professor to allow them use of the relevant facilities and to give them guidance. Taylor described the procedure in an address at the Semicentennial Celebrations of Rice University more than 50 years later (1963c):

> When I had taken my first degree after three years at Cambridge, I asked my professor at the Cavendish Laboratory, J.J. Thomson, whether he could suggest a research project which I might attempt. This was the regular way for students to begin their careers, and I think few of us stopped to think of the amazing fertility of a mind which could produce a whole string of worthwhile ideas for each of a dozen or more students. J.J. produced a number of attractive ideas, and I chose one of them. It was this: If, as had recently been suggested, light consists of spots or quanta of energy localized in space, interference must depend on the phase of a quantum falling on any particular place being somehow related to the phase of another quantum falling later on the same spot. If, then, the intensity of light forming an interference pattern is reduced, the time interval between the incidence of successive quanta on the same spot must increase, and, if the intensity is reduced to an exceedingly low value, the phase of the disturbance produced in a photographic grain by a quantum may not remain unimpaired for a sufficiently long time while waiting for the next quantum to interfere with it. J.J. therefore suggested that I might set up a simple interference pattern and see whether it ceased to be well defined when the light intensity was reduced.
>
> I chose that project for reasons which, I fear, had nothing to do with its scientific merits, and set up in the old children's playroom at my parents' home a vertical needle on to which I shone the light of a gas jet placed behind a vertical slit made by a

razor blade in a piece of metal foil stuck to a piece of glass. I set
a photographic plate up some six feet away and obtained a good
interference pattern. Then I reduced the light by inserting
successive sheets of dark glass between the gas jet and the slit and
increased the exposure time. I got to the stage when I calculated
that I would get enough blackening of the plate if I made the
exposure six weeks, and I had, I think rather skillfully, arranged
that this stage would be reached about the time when I hoped to
start a month's cruise in a little sailing yacht I had recently
purchased.

With maximum darkening the amount of energy falling on the
photographic plate was the same as that due to a standard candle burn-
ing at a distance of about one mile, but after three months exposure
the plate showed a pattern which was as sharp as those shown in figure
4.1. Thus this experiment, performed with home-made apparatus,
was successful in giving a definite negative result which is referred to
in books on optics.[1]

It is remarkable that Taylor did not get further involved in the
exciting new ideas that were transforming physics at the time and
that were no doubt being discussed vigorously in the Cavendish.[2]
Later in the same address (1963c) he said 'I did not feel a call to a
career in pure physics.' I think it likely that his instinctive feeling
of independence made a fashionable field seem less attractive, but it
must have required a good deal of self-confidence and courage for
him to go his own way at the age of 22.

1. I am told that on a more modern understanding of quantum theory it
   was not to be expected that interference fringes would cease to form,
   however small the intensity of the incident light may be. The distinguished physicist Richard Feynman described this and other similar
   results as the 'central mystery' of quantum physics.
2. There was another piece of research in physics which is not mentioned
   by Taylor in any of his later writings and which was apparently not
   published. Soon after his death in 1975 a water-stained incomplete
   manuscript describing an experimental investigation of the action of
   crystal rectifiers was found in his garage. The paper records that the
   work was done in response to a suggestion made by J. J. Thomson
   'some months ago', and this provides the only evidence of its date.

(a)

(b) (c)

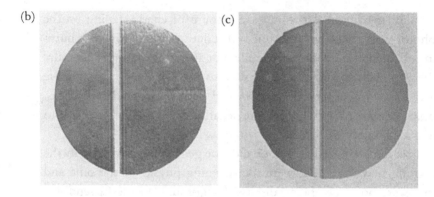

Figure 4.1 *A test of the new quantum theory; do interference fringes continue to form when the intensity of the light is greatly reduced? (a) The apparatus; (b) exposure time 4 minutes; (c) exposure time 24 hours. (From 1909a)*

Taylor's second published paper (1910a) – on the structure of a shock wave in a gas – arose out of his own reading, perhaps of a note by Rayleigh published a little earlier.[3] Taylor established theoretically that a propagating sharp transition layer of permanent type (that is, one which does not change its form as it propagates) is possible provided two conditions are satisfied. One of these conditions is that the pressure must increase in the direction in which gas flows through the transition layer. This has the consequence that nonlinear effects cause the pressure gradients to increase in magnitude within the layer and the transition to become steeper, whereas if the pressure

3. *Proc. Roy. Soc. A* **81**, 1908, p.449.

decreases in the direction in which gas flows through the layer nonlinear effects cause the transition to become less steep. The other condition is that diffusion processes must operate in the interior of the layer in order to balance the convective steepening and yield the desired steady state relative to the shock. He then set up the continuum equations for the velocity, density and temperature in the interior of the propagating layer for a perfect gas with constant viscosity μ and heat conductivity κ, and showed that they could be solved exactly if either $\mu = 0$ or $\kappa = 0$ and approximately if the velocity jump across the layer is relatively small. The solution in this latter case makes clear the way in which the nonlinear convection terms in the equations of motion cause over-turning of the wave, and it gives an order-of-magnitude estimate of the thickness of a shock wave, viz. the kinematic viscosity of the gas divided by the velocity jump. This approximate solution is particularly valuable, because the continuum equations may be expected to be valid (approximately) only when gradients of the gas velocity are small. The idea of exploiting the analytical simplifications that obtain when μ and κ have certain specific values or when the velocity jump is small is characteristic of Taylor; he found it easier and more profitable to think about concrete cases, and he knew intuitively when the assumed parameter values were untypical or misleading.

This mature paper in which a basic phenomenon in gas dynamics was clarified was Taylor's debut in fluid mechanics, the field in which he was to publish over 150 papers during the subsequent 60 years. He was awarded the Smith's Prize[4] for his work on shock waves. More importantly he was elected in 1910 to a Prize Fellowship at Trinity College which provided support and freedom to pursue his research for up to six years.

Taylor made massive contributions to many broad areas of fluid mechanics – turbulent diffusion and dispersion, the dynamical structure of turbulence, dynamical meteorology and oceanography, aerodynamics, the mechanics of explosions, low-Reynolds-number hydrodynamics, and others – and it is natural to suppose that he

---

4. A competitive prize for original work by a senior mathematics student at Cambridge.

chose deliberately to work in those areas because he saw that at the time they had promise, and offered opportunities for contributions of the kind he could make. However, the evidence suggests otherwise. Taylor himself said that 'the course of my scientific career has been almost entirely directed by external circumstances'[5] and that he simply reacted to events.

A major external development early in his research career to which Taylor responded and which influenced his research interests for many subsequent years was an opening in meteorology. In 1907 the Meteorological Committee, a national government body, was given money by Arthur Schuster which would enable them to appoint a Reader in Dynamical Meteorology for a period of three years. Schuster was a wealthy man who was Professor of Physics at Manchester University, and his purpose was to encourage a mathematician to study dynamical meteorology and to introduce some analytical and quantitative reasoning into what was then a rather empirical subject. The Readership could be held at any British university. The first holder was E. Gold, who chose to hold the Readership at Cambridge and spent his time on a pioneering investigation of radiative equilibrium in the stratosphere. Schuster declared in 1910 that he was pleased with the success of the experiment, and made another donation to provide for a second appointment for a similar period of three years. This time Taylor was an applicant for the position, although I do not think he had previously formed any intention of taking up meteorology, and he was appointed to begin on 1 January 1912, soon after recovery from a severe bout of pleurisy which confined him to a sanatorium for several months. He too chose to hold the Readership in Cambridge. It was a bold but quite inspired decision by the electors to appoint this eager young man who, like Gold before him, was still in his mid-twenties on appointment. Dynamical meteorology proved to be an ideal field for him, for it gave him scope for exercise

5. G.K. Batchelor, An unfinished dialogue with G.I. Taylor, *J. Fluid Mech.* **70**, 1975, pp. 625-638.

of the adventurous curiosity that I think was the main driving force of his life.[6]

Taylor spent much of 1912 reading about dynamical meteorology, giving lectures on this topic, and thinking about suitable areas of research. Although he did not publish any papers in 1912, there are later references to work of an exploratory kind begun during this period. The description of various models of the general circulation that had recently been published by Cleveland Abbé did not attract him, because it seemed to be difficult to connect the analytical results of any of these models with quantities which could be measured. He saw more promise in investigations of small-scale processes, and became interested in the mixing and vertical transfer of momentum and heat that occurs in the lowest layers of the atmosphere as a consequence of turbulent velocity fluctuations. One of his first experiments concerned the directional properties of these velocity fluctuations near the ground. He made a light double-jointed wind vane capable of recording vertical as well as transverse components of velocity, and found that, whereas at 1 or 2 ft above flat ground the transverse fluctuations were (statistically) much larger than those in the vertical direction, they were nearly equal at heights of 10 ft or more. Recordings of a vane of this type were published later (1927e); see figure 4.2. He also found, from a comparison of the widths of the wind-direction spread and the wind-strength spread recorded by a wind vane on a tall chimney, that the velocity fluctuations in the wind direction are approximately equal to those in the transverse direction, showing that except very near the ground the turbulent velocity fluctuations are roughly isotropic (1917d). This simple conclusion was at variance with the then common conception of turbulent eddies in a shearing motion as being like rollers causing wind fluctuations

6. The minutes of a meeting of the Meteorological Committee in March 1915 contain a reference to a Professorship of Meteorology to be financed by the War Office and to be held by Taylor in association with the Royal Flying Corps. There are no records of Taylor's work specifically in this capacity, although it is known that in 1917 he was appointed as meteorological adviser to the Royal Flying Corps.

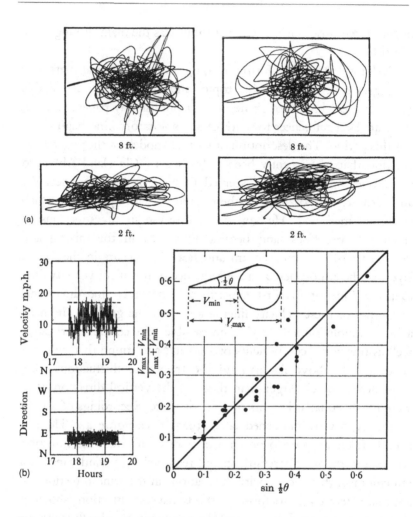

Figure 4.2 *(a) Records of the vertical and transverse components of the wind velocity obtained with a double-jointed wind vane. (b) Records of the magnitude and direction of the wind velocity on a tall chimney on 8 July 1917. The limits of variation of records like these are used in the right-hand graph to show that the turbulence is approximately isotropic. (From 1917d and 1927e)*

mainly in the vertical plane parallel to the mean motion. Crude though the observations were, they stimulated him to begin the thinking about the nature of turbulence that was to be a major pre-occupation for the next 21 years.

Taylor realized for instance (1917d) that turbulent motion *must* be three-dimensional, for consistency with the common observation that the rate of dissipation of kinetic energy per unit volume of fluid remains finite as the viscosity is allowed to become small.[7] The rate of dissipation in the volume V of the fluid is

$$\int \mu \omega^2 dV,$$

where $\omega(= \nabla \times \mathbf{u})$ is the fluid vorticity and μ is the viscosity; and the equations of motion show that when the motion is two-dimensional (so that $\mathbf{u}.\omega = 0$ everywhere) the vorticity in a material element of fluid is constant when μ is small. But when the motion is three-dimensional, the vorticity magnitude $|\omega|$ may be increased greatly by extension of vortex lines and the rate of dissipation may remain finite as $\mu \to 0$. He went on to argue that 'the eddy motion must tend to produce small whirls or discontinuities where a very large vorticity occurs in a very small volume.' He is absolutely correct, and about 30 years ahead of others, in concluding that the spatial distribution of vorticity has a spotty or intermittent character, although he appears to be using his intuition, rather than mathematics, when he rejects the possibility that sufficient largeness of the vorticity magnitude is brought about by extension of vortex lines everywhere in the fluid and not only in spots.

Taylor's preparations for more accurate observation of the wind fluctuations near the ground did not bear immediate fruit, because in 1912 there was another of those external events to which he responded. As it turned out, this new development gave him scope for more ambitious and exciting observations of the rate of transfer of momentum, heat and water vapour in the friction layer of the atmosphere over the sea. It was consequently not a change of direction, more a change of scale of the enquiry.

At all events, Taylor's first steps in research were over. He had become a master rather than a learner, as the impressive work described in the next chapter makes clear.

---

7. This could well be regarded as the fundamental law of turbulence.

CHAPTER 5

The Scotia expedition

On its maiden voyage in April 1912 the s.s. *Titanic* collided with an iceberg in the North Atlantic and sank with heavy loss of life. This disaster led the British Government and some shipping companies to equip an expedition on an old 230-ton wooden whaling ship, the s.s. *Scotia* (figure 5.1). The purpose was to look for icebergs in the North Atlantic and report their positions by the new radio method. Three scientists were included in the expedition, a meteorologist, an oceanographer and a plankton specialist. Taylor was invited to be the meteorologist and spent roughly the first half of 1913 at sea on the *Scotia*. It may be that in accepting the appointment he was motivated mainly by his spirit of adventure, but in the event it proved also to be a scientifically fruitful period. His

Figure 5.1 *The s.s. Scotia, an old 230-ton whaling ship.*

duties included the flying of kites and balloons fitted with meteorological instruments to measure and record the air temperature and humidity in and above the fogs which are prevalent on the banks of Newfoundland (figure 5.2).

(a)

(b)

Figure 5.2 *The aft quarter-deck of the Scotia from which the balloons and kites carrying instruments were launched. (a) Taylor took this photograph from the ice in which the ship was held; (b) the first mate and second engineer;*

(c)　　　　　　　　　　　　　(d)

Figure 5.2 *(cont.) (c) a tethered balloon; (d) a kite ready to be flown from the head of the rear mast.*

Getting the kites up was not straightforward, as Taylor described later in his Hitchcock Lecture at the University of California (1952d):

> This was a very different kind of work from the formal mathematical analysis which I had studied as an undergraduate. The first requirement was tact in dealing with the *Scotia*'s captain. The *Scotia* was a wooden barque-rigged sailing ship of a type which is now extinct but which had formerly been used by whaling expeditions when whales were attacked by hand harpoons from ships' boats. She had single topsails which made her look rather like the *Mayflower*, but these were easily reefed by rolling them up on a thing like a roller blind attached to the underside of the topsail yard. She had a 40 horse-power auxiliary engine which was incredibly inefficient but would drive her at about $4\frac{1}{2}$ knots in a calm. The *Scotia* had been used by the Scottish explorer Bruce in the Antarctic and Bruce had tried to fly kites from her in much the same way that children fly them in a field. He had a whaleboat lowered in order to carry the kite downwind. When the boat was a few hundred yards downwind of the ship a sailor in the boat released the kite while men on the ship pulled on

the kite line. Every time this manoeuvre was attempted the kite would rise a little and then gyrate wildly and dive into the sea. The *Scotia*'s captain, who had also been the captain during Bruce's expedition, told me that Bruce had never succeeded in raising a kite from the *Scotia* and that the waste of time and effort during these attempts to do so were so great that he would refuse to command the ship if kites were to be used.

It was my job not only to find out how to fly kites from the ship but to persuade the captain that it could be done. I guessed that the reason Bruce's kites misbehaved was that he had not been able to get them out of the eddies which lie downwind of any structure and might be expected to be very violent to leeward of a sailing ship because of its big windage. To enable the kite to get into the undisturbed air above the ship I arranged to fly it from a block at the top of the mizzenmast. The kite cord, which was a steel piano wire, led from a special winch on deck up to this block and down to the kite which was held by two seamen standing on the quarterdeck. On the word 'go' the winch started and the kite was lifted upwards out of the sailors' grasp. It gyrated wildly as it blew out downwind but of course it could not get into the sea because it was hanging from so high a point. After a few gyrations it always got up into the undisturbed air above the top of the mast and stayed there flying steadily. In this way I got the kites up to 6,000 feet. The captain raised no objection to kite flying when he saw them in the air, but when the ship was being fitted with a kite winch in Dundee a mild deception had to be practised. The fact that soundings were to be made in the air and not in the sea was not mentioned.

The method I have described for getting the kites into the air always worked but it was not reversible. When the kite was wound down to the deck it would begin to gyrate as soon as it got down to the level of our sails and rigging. It usually dived into the sea from which it would easily be recovered. After such an immersion I used to take out the clock which drove the recording drum of my meteorological instrument, empty it of any sea water, dip it in several changes of fresh water and then heat it up before the cabin stove. It always survived this treatment undamaged.

While away on the expedition Taylor wrote a number of letters to his parents. Here is a revealing one to his father in which some of the difficulties faced by a meteorological observer at sea are described and the mood is less light-hearted than in the much later reminiscences:

s.s. *Scotia*, c/o Furness Withy & Co,
St John's, Newfoundland. June 3rd, 1913

Dear Father,

We arrived here late on Saturday night, and on Sunday morning I had several letters one of which was from Mother. This afternoon they brought me some registered ones from you and Mother. I was awfully glad to hear about the Wheelwright's Shop.[1] I hope it was not put near any large or very brightly coloured pictures. Will it appear in the yellow book on every gentleman's drawing-room table? Yours was not the only success in the artistic world that I heard about in my letters. Adrian sold some of his pictures, but I will tell you about that afterwards[2] as I expect you will want to hear about us.

We left here on April 23rd and went down South East till we got into the Gulf Stream. We then turned NE and afterwards zig-zagged across the iceberg region gradually making northwards till we got to the pack-ice off Belle Isle Strait. We then pushed along for several days forcing a passage through heavy pack till we were about the latitude of Hamilton inlet, Labrador. After that we went SE again and finally came in here.

We have been going the whole time and I have never been allowed to have the ship at my disposal for scientific work, in fact if I had known that we were to spend our whole time ice scouting I should certainly never have come. As time went on and I had no opportunity of doing kite or balloon work I decided to try and make kite ascents while the ship was going and at the risk of losing apparatus. I was, as you know, engaged for a 3 months cruise which might possibly extend into the 4th month, but would certainly not last longer. The instructions to Capt Robertson stated that we should be sent North after our second call at St John's *if there was time*. It is now after the time we were due back in England so I naturally supposed that we should return home. I therefore sent up kites on every possible occasion and consequently lost a good number of them. Now that it turns out that we are going to be away for goodness knows how long, this is a great pity.

1. One of Edward Taylor's paintings.
2. Unfortunately, he did not do so, at least not in a letter which has survived. Edgar Adrian, later Lord Adrian OM PRS and a distinguished physiologist, was a life-long friend of Geoffrey Taylor who in collaboration with some contemporaries at Cambridge perpetrated a famous hoax on the British art world in 1913 by painting and exhibiting a number of pictures in what they described as the 'post-Impressionist' style.

I am busy getting repairs done at present, but I shall not know how much apparatus I shall have to go on with till the end of the week. I have had a great deal of difficulty with my apparatus, partly owing to the fact that the *Scotia* is so unsuitable a craft for meteorological observation and partly owing to the haste with which the the expedition was got up.

I expect you will have seen a letter from St John's in which the *Scotia* was criticised and in which we were made responsible for the criticisms. I should like to meet the man who wrote it and throw him overboard. The letter was dated April 16th, 2 days after we arrived here. I had spent my whole time during those days in going on walks alone into the country while Matthews and Crawshay[3] tell me they certainly gave no information whatever that could possibly lead anyone to suppose that they harboured such thoughts about the *Scotia*. The writer of the letter was evidently a party interested in the St John's sealing ships and in running down the *Scotia*. He made use of our names to give his criticism of the scientific work some authority. This method comes out most strongly in his remarks on the wireless installation because he says the Marconi operators criticise the installation. At the time the letter was written the operators had not even been ashore and no one had come aboard except the agent. They say they had not met him at all even then.

The inner history of the affair appears to be this – as far as we can make out. Many months ago the B. of T. approached some people in St John's with a view to finding out what suitable vessels could be chartered there. They finally decided on a vessel from Dundee, St John's traditional rival and enemy. The interested parties in St John's then got up this case against the *Scotia* and wrote the article in which correct and incorrect facts were ingeniously mixed.

. . . .

As I have said before I have had hardly any opportunity of doing any work besides the regular daily observations and getting the Barnes self-recording thermometer working. I had a good deal of trouble with this last. As supplied it was quite unsuitable for the work so I had to make extensive alterations. I had to risk spoiling the whole affair by sawing it right round. Luckily I managed to do this without breaking the platinum wire of the thermometer. After drying it out (it had got sea water into it) I made the mouth watertight, insulating with some Chatterton's compound borrowed from the Marconi operators. I then

3. The oceanographer and the plankton specialist.

managed to make a joint round the saw-cut by soldering on a strip of an old beef tin. This was really a difficult job because if the insulating paper inside got the least bit burned the instrument would be spoiled so I had to use the utmost circumspection.

I managed to get the job done all right and took records of sea temperature but I found that Barnes' results[4] certainly do not apply to the region we were working over. Barnes worked in the mouth of the St Lawrence where the water is more or less homogeneous. We worked in the junction of the Gulf Stream and the cold Labrador current where the water is so variable in temperature that it may change 4°C in less than a mile. Barnes' maximum temperature change due to icebergs was about $3\frac{1}{2}$°C in 4 or 5 miles. During the next part of our cruise we shall be in some more homogeneous water, so I shall have a better opportunity of testing Barnes' results, but my work up to the present shows, I think, that in the regions where icebergs are most dangerous to shipping, namely on the steamer routes, the Barnes apparatus would be no good for detecting the presence of bergs.

I have sent up a number of kites and have got very few results owing to various causes, chiefly of course because we have always been making passages and all scientific work has been subsidiary to that. Once I sent up a kite in a fog. It came down owing to a lull in the wind and fell into the sea. I succeeded in persuading the Captain to stop the ship for 20 minutes while I wound in the wire. The kite was spoiled; so was the mainspring of the clock on the instrument. The card on which the results were recorded was made of good stuff and by careful drying I was able to read it. The results were quite the most interesting I have had yet. The temperature was nearly freezing at the surface and about 60°F a thousand feet up. I can't help thinking that this result may give a clue to the reason for the permanence of certain types of fog which occur on the Banks.

I have not had any chances of using the balloons. I think that by perseverence and seizing every opportunity I may get some results worth having but I wish I had never come. The worst feature of the ship from my point of view is the Captain. He is very stupid and his attempts at economy would seem mean if they were not so funny. We

4. In his report on the work of the expedition (1914a), Taylor referred to an observational result found by H.T. Barnes that the surface water is a little warmer in the vicinity of an iceberg. Taylor made careful measurements of the sea temperature as the *Scotia* approached and passed an iceberg, and could find no significant effect of the presence of the iceberg. Nor did he find Barnes' explanation of how such an effect might be produced convincing.

have electric light aboard and the Captain is always inveighing against it, saying that he has been forced to give up writing or reading in his cabin because of it. One day he was telling me this yarn and he added that he infinitely preferred a candle, that he was accustomed to reading and writing by candle light. I asked him why he did not use a candle then and he announced in a tone of pained surprise 'What, use a good candle when the dynamo will be kept going in any case!'

This sort of thing would be all very well if he kept it to himself but unfortunately he applies it to our food. You will see from this how used I am getting to sea customs – every true sailor grumbles at his food. Capt Robertson's chief redeeming feature is his kind heart; he is well meaning, but oh so stupid!

I think I had better stop here partly in consideration for your eyes, but chiefly because if I go on much longer there will be nothing left to say when I write to her. By the bye I enjoyed all the letters you sent along enormously. You need never want for something to say because the most unimportant little bits of home news assume a quite surprising interest over here.

Love from Geoffrey.

Taylor's scientific work on the expedition was motivated by his wish to be able to describe common turbulent processes quantitatively. In the case of the turbulent flow in the friction layer of the atmosphere, an important process is the vertical transfer of transportable quantities such as heat and water vapour. Taylor noted that the air temperature at low levels is usually not very different from that of the sea, and wrote (1914a): ' It seems probable that the heating or cooling effect of the sea will extend upwards through the atmosphere and will do so at a rate which might be expected to depend chiefly on the wind velocity. In analysing the data obtained from the kite ascents this idea has been kept in view.' By writing the local potential temperature of the air as the sum of a mean over a horizontal plane and a fluctuation about that mean, Taylor showed by heuristic arguments that as a consequence of heat conservation the mean potential temperature satisfies the familiar diffusion equation, in which the diffusivity is of the general form $\langle |w| \rangle l$. Here $\langle |w| \rangle$ is the mean magnitude of the fluctuation in the vertical component of the velocity of the air and l is the average vertical distance travelled by an element of fluid before it acquires a different temperature by mixing with surrounding fluid. (In the later terminology, l is the 'mixing length'.) The point of Taylor's quoted remark is that this 'eddy conductivity' may be expected to depend mainly on the mean wind speed, in which case empirical estimates should be grouped according to the values of that wind speed.

The sponsors of the *Scotia* expedition required Taylor to measure the vertical distributions of mean temperature, humidity and wind velocity at different locations because these quantities are presumably relevant both to fog formation and the incidence of icebergs, and the results were described in a formal report (1914a). Taylor also used the observations for the more fundamental purpose of inferring the values of the eddy conductivities for heat and water vapour, essentially by comparing the observed speed of vertical propagation of a jump or region of rapid change of mean gradient with what would be expected from the diffusion equation (1915a). There is however a difficulty about such a procedure, namely that the sea temperature at the position of the ship at the time of the observations is not necessarily the same as the sea temperature at the position of the same air mass at a previous time. An observer on the ship can interpret the measured vertical distribution of potential temperature in a particular air mass at a particular time only if he is given the sea temperature beneath that same (moving) air mass at previous times. Taylor met this difficulty by making use of the records of sea temperature and wind velocity at bridge level compiled by ships in the general neighbourhood of *Scotia*. The records were not available to Taylor until he returned to London, so for lack of records of the sea temperature and wind velocity at the right spot at the right time he was unable to make use of many of his vertical traverses. In the event he was able to construct the previous air path and sea surface temperature for seven of the traverses.

As an example, figure 5.3 shows the previous positions of the air mass in which the vertical distributions of temperature and humidity were observed at 7 p.m. on 4 August 1913. Isotherms of the sea surface, assumed to be constant over an interval of several days, are also indicated in figure 5.3, and figure 5.4 shows the results of the vertical traverse of the air mass by the ascent of a captive balloon. The region from the sea surface up to about 370 metres in which the vertical gradient of mean potential temperature is approximately uniform and

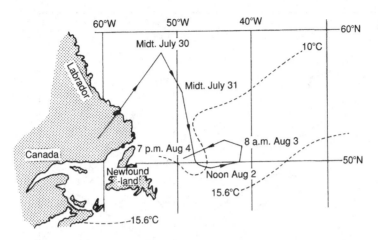

Figure 5.3 *Isotherms of the sea temperature, and the path of the surface air mass ending at the Scotia at 7 p.m. on 4 August 1913. (From 1915a)*

positive evidently represents the effect of cooling of the base of the air mass during the preceding 36 hours in which the air mass was moving westward into the cold water of the Great Bank of Newfoundland. Other features of the temperature distribution can be interpreted similarly, and the mean humidity distribution in figure 5.4 shows that changes in the water-vapour content of the air are propagated upwards in the same general way as changes in the temperature. By comparing the speed of upward propagation of kinks in the temperature distribution with what would be expected if the temperature satisfied a diffusion equation, Taylor found the effective diffusivity did not vary much over heights up to 770 metres, and was roughly 3×10^3 cm^2/sec for wind forces of about 3 on the Beaufort scale.

Taylor next considered the vertical transfer of fluid momentum when the horizontal component of the mean wind velocity depends only on vertical position. He perceived that the momentum of an element of fluid cannot be supposed to remain constant over a certain interval and then to change by mixing with the local fluid with overall conservation, as happens in the case of temperature and humidity, because fluctuating pressure gradients affect the momentum of the element. However, in a fully two-dimensional turbulent motion, the vorticity of a fluid element is conserved (aside from the small effect

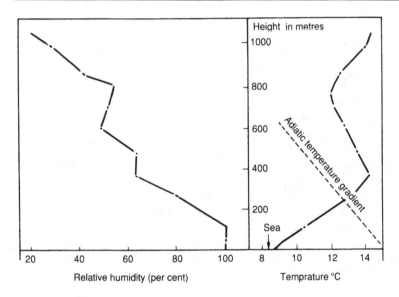

Figure 5.4 *The vertical distributions of relative humidity and temperature measured at the position of the Scotia at 7 p.m. on 4 August 1913. Weather: thick fog; wind speed about 5 m.p.h. at all heights. (From 1915a)*

of molecular viscosity), in which case the same argument as for temperature shows that the mean vorticity satisfies a diffusion equation in which the eddy diffusivity, or eddy viscosity as it is now, is again $\langle |w| \rangle l$. Taylor had no means of estimating the errors introduced by this unjustified assumption of two-dimensional motion and set the question aside until suitable observations became available.

The last section of this pioneering paper on atmospheric turbulence (1915a) was about the effect of the Earth's rotation on the vertical distribution of mean velocity. If there were no frictional stress at the Earth's surface, the air velocity at all heights would have the geostrophic value such that the horizontal pressure gradient balances the Coriolis force. However, when there are frictional stresses across horizontal planes, the air velocity has a low value near the surface and increases with height, finally asymptoting to the geostrophic value at a height which varies between 200 and 1000 metres. This is the so-called friction layer of the atmosphere. It is easy to see that in the steady state both the magnitude and direction of the wind velocity must vary with height, and it is not difficult to formulate the equation of motion for a case in which the wind velocity depends on height alone and the stress across a horizontal plane is proportional to the local mean velocity gradient.[5]

Unknown to Taylor, Ekman had done this in 1905[6] and had solved the equations for steady laminar flow with molecular viscosity. As a boundary condition Ekman assumed a given value of the tangential stress at the boundary, and he applied the results to wind-generated ocean currents on a rotating earth. Taylor obtained the same general solution, but he regarded the viscosity as being an 'eddy viscosity' representing the effect of turbulent velocity fluctuations (taken as being statistically independent of height for convenience) and in place of Ekman's boundary condition he assumed the velocity at the boundary to be proportional to the stress there, as is appropriate for turbulent flow over a rough rigid surface. The fluid velocity is found to vary with height in a spiral fashion, and the angle between the surface wind and the geostrophic velocity varies with the wind strength, being about 20° for light to moderate winds; and the air velocity is close to its asymptotic value at a height which is proportional to the square root of the eddy viscosity. Taylor found general agreement between these theoretical results and balloon observations of wind velocity which had recently been made by Dobson[7] over Salisbury Plain, especially when empirical values of the eddy viscosity and related quantities were grouped so as to refer to a given small range of values of the geostrophic wind velocity.

For good measure Taylor added in an appendix to his paper (1915a) the penetrating suggestion that, since in a turbulent shear flow momentum is transferred to the boundary at a rate which is practically independent of the molecular viscosity, Rayleigh's recent demonstration that instability of steady flow of inviscid fluid between parallel plane rigid boundaries is impossible when the second derivative of

5. *An Introduction to Fluid Dynamics*, §4.4, by G.K. Batchelor. Cambridge University Press, 1967.
6. V.W. Ekman, *Ark. Math. Astr. och. Fys*, **2**, no. 11, 1905.
7. G.M.B. Dobson, *J. Roy. Met. Soc.* **40**, 1914, p. 123.

the undisturbed velocity is one-signed will not necessarily apply to a fluid of small but non-zero viscosity.

Taylor's work on the turbulent friction layer of the atmosphere, carried out when he was 27, shows that as well as being outstandingly original he was capable of organizing comparatively large-scale experimental resources and did not hesitate to do so when the scientific purpose on hand required that. It would be a mistake to suppose that the essential characteristics of his research are incompatible with or inappropriate for the use of large-scale facilities. It is true that during that long and very fruitful post-World War II period in his sixties and seventies, which is now the only period of his life of which we have much first-hand knowledge, his lifelong taste for independence led him to choose problems which could be studied experimentally with no resources other than a single technical assistant, a one-room laboratory and simple apparatus. We recognize insight, simplicity and ingenuity in the small-scale experiments of this period, and no doubt the purest examples of these qualities are to be found in these experiments, but they are also displayed in the large-scale projects undertaken by Taylor in his earlier years. One of the people invited to the Symposium on 'Fluid Mechanics in the Spirit of G.I Taylor' in March 1986, Carl Friehe at the University of California at Irvine, wrote to say that unfortunately he could not attend because he would be involved in setting up a Navy aircraft to measure wind velocity, turbulence and pressure gradient in the lower atmosphere. He added that he had reflected many times on G.I.'s kite-borne measurements from a ship 73 years earlier, because 'with all of our modern equipment and its headaches it is not clear we will add anything more to the understanding of atmospheric turbulence than G.I. did with his incisive experiment.'

This is a suitable place for comment on an amusing foible which stayed with Taylor all his life, and which showed that he too was human. Most of his experiments were done to test some idea or calculation after he felt he understood what was going on rather than to acquire data from which understanding would come later. He usually knew in advance what to expect and was seeking a measurement for comparison with some theoretical prediction. Because his insight was so acute, it seldom happened that the experiment failed to yield the expected result; and if there was some lack of numerical agreement it could often be put down to the effect of some extraneous factor. The form of experimental check that he liked best was the observation of a single number of which a theoretical value was already available, and I think he sometimes stopped making measurements as soon as he had obtained fair agreement with the theoretical value, with a blithe lack of concern for the rules about the number of significant figures and the theory of errors. Sometimes his conscience seems to have pricked him, although not to the point of compelling stern action, for in a number of his papers there is a gleeful admission that the agreement obtained is better than could be justified by the accuracy of the measurements and the nature of the theory.

Taylor's comparison of some balloon observations by Dobson of the dependence of the mean wind velocity on height above Salisbury Plain with the theoretical spiral distribution provides the first example of this strange lacuna. Dobson's observations gave the heights at which the wind velocity first reaches the same direction and the same magnitude as the geostrophic wind as 800 m and 300 m respectively for light winds, the estimates being crude because locating the point of intersection of two close curves is intrinsically inaccurate. Moreover the calculated ratio of these two heights (viz. 2.6) is a mean over several quite different values for different wind strengths. Taylor concludes (1915a, p. 18) that 'it is probably a coincidence that the observed ratio of these two heights, viz. 2.66 (= 800/300), should be so very close to the calculated ratio 2.6, but the coincidence is at least significant.' What can he mean by 'probably' and by 'significant'? Or see his paper on the blast wave from an intense explosion (1950d), where he estimates the velocity of rise of the luminous hemisphere created by the first atomic bomb explosion by regarding it as a vacuous bubble of radius equal to that where the density of the heated air is half that of the surrounding cold air – a half being the average of one and zero – and ends up with an estimate which differs by only 2 per cent from the observed value, which itself is uncertain to within a much bigger range, the conclusion being 'It will be seen that the agreement is better than the nature of the measurements would justify one in expecting'. There are other examples. Taylor was seldom in doubt about the general validity of his theory, and I am sure that any cavalier treatment of observations never led him to incorrect conclusions. I believe his enjoyment of the process of finding agreement between theory and observation was so intense that he was sometimes led unconsciously to 'help' the agreement a little in ways which we would normally regard as inadmissable.

I got my fingers burnt once by assuming that he too saw the humorous side of what he was doing. I was acting as chairman of a seminar lecture which he had just given in the Cavendish, around 1950. Thinking to raise a laugh I cheekily remarked on Taylor's uncanny ability to get good agreement between his theory and experiment. He did not like it at all, and I do not know who was the more embarrassed, he or I.

Taylor would be entitled to claim that he was in good company. There is a story about Rutherford – another physicist who did not need calculations to help him to understand physical phenomena – that he was once calculating the atomic mass of an α-particle with the help of a slide-rule during a lecture, and said 'we find 3.4, or, on rounding this off to the nearest integer, 4'! I mentioned Taylor's usual way of comparing theory and observation to Teddy Bullard, a distinguished geophysicist, and his comment was that, with the notable exception of Harold Jeffreys, no physicist of Taylor's generation paid any attention to statistics.

CHAPTER 6

Participation in the birth of aeronautics

The exciting possibilities of powered flight were evident in 1914, and it was almost inevitable that Taylor should get drawn into the development of aeronautics and its use for military purposes. He spent some happy years in a group of civilian scientists, with a variety of backgrounds, at or near the Royal Aircraft Factory at Farnborough, and, as could have been predicted, he learnt to fly and to conduct his own in-flight experiments. This was for him an ideal kind of activity, with appeal to his spirit of adventure and with scope for exercise of his technical ingenuity. Later he reflected: 'I am glad to have lived at a time when they (aeroplanes) were simple enough for a reasonably intelligent amateur to appreciate them as mechanisms.'

The early days of aeronautics became Taylor's favourite topic for a lecture of reminiscences to a scientific audience in his later years, and he gave several such lectures with slightly different titles and contents, most of which have been published (see 1966f, 1969e, 1971b). His style in these lectures and articles is anecdotal and light-hearted, which suits this particular subject (although the sharpness of his intellect is sometimes concealed by the levity). The remainder of this chapter is a reproduction of the typescript, hitherto unpublished, of a lecture given first as the Lester Gardner Lecture at MIT in May 1971 (when he was 85) under the title 'Aeronautical experience before 1919', and second to the members of Trinity College, Cambridge, in October 1971, with the title, 'Aeronautics before 1920'. It is a piece of the history of technology, described entertainingly by one of the leading participants.

The first time I saw an airplane was in 1912. I was visiting a friend in Lancing College when I looked out from a window towards the airfield at Shoreham about two miles away and saw the start of one section of a race round Britain. I saw the eventual

63

winners, two Frenchmen, Messieurs Beaumont and Vedrines, take off and fly up the shallow valley on one side of which Lancing College is built. Although I was only about 200 feet above sea level I looked *down* on the top of the plane as it passed.

My active participation in aeronautics was accidental. I had studied fluid dynamics as an undergraduate at Cambridge and was applying my knowledge to meteorology when the first world war broke out on August 4th, 1914. Like most of my friends I thought I ought to do something about it, so I went up to London the next day and volunteered for service in the Army. I suggested to the War Office that if I could be given telegraphic facilities I might set up a weather forecasting service for the army in the field. The officer to whom I made this suggestion did not seem to doubt that I could tell exactly what the weather was going to be – as he might very well have done – but thought the knowledge would be of no value to the army in the field. 'Soldiers don't go into battle under umbrellas, they go whether it is raining or not' was his comment. I expected then to have (I must confess unwillingly) to join a combatant unit but I left my name and my parents' home address in London with the officer I had spoken to in case the army had any use for my service as a scientist. Shortly after my interview this officer happened to meet Major Sefton Brancker who was at that time, before the creation of the Royal Flying Corps, in command of the flying wing of the army. He told Brancker of my visit and asked if he had any use for scientists in his department. Brancker replied that he could use as many as he could lay his hands on, and that evening he sent a messenger to tell me that the Royal Aircraft Factory at Farnborough would employ me as a scientist if I would offer my service in that capacity.

I went there the next day and the Superintendent Mervyn O'Gorman introduced me to a fascinating character, Teddy Busk, a King's man who had graduated at the top of the Engineering Tripos list in 1907. Teddy invited me to live in the house which he shared with the Naval Airship mess. At first only Busk and I and F.M. Green were not in the Navy, but later we were joined by other civilians, F.W. Aston, the inventor of the mass spectroscope and the principal discoverer of isotopes, and Lindemann, afterwards Lord Cherwell. The mess subsequently moved to a house called 'Chudleigh'; I cannot remember who were members at the time of the move, but Farren, Glauert, R.H. Fowler, G.P. Thomson, E.D. Adrian and Melvill Jones were certainly 'Chudleighites'.

Busk was perhaps the first man to make a practical study of the stability, as distinct from the controllability, of airplanes. The machine, BE2C, which he had redesigned to be stable, could be flown without hand control, and a thorough study of its aerodynamic character was being made when I began work at Farnborough. Longitudinal stability was achieved by enlarging the tail plane of the previous BE models and ensuring that the centre of gravity was correctly situated in relation to the wings. Lateral stability was ensured by giving the wings a dihedral angle, that is by sloping them slightly upwards from the body.

My first flight in an airplane was with Teddy Busk on November 4th 1914 in the prototype BE2C in which he was timing longitudinal oscillations. To produce these he pushed the control column forward so that the machine dived and the speed increased considerably. Then he let go and the machine executed a series of oscillations which were recorded by instruments. The next day he was continuing his experiments on the same machine when it caught fire in the air and he was killed. During my flight with him I had noticed a strong smell of gasoline and he had explained that the carburettor was supplied from a small gravity feed tank in front of the passenger's seat. At frequent intervals this had to be filled by the pilot who pumped petrol into it from a large tank under his seat. Unfortunately there was no indicator to show the pilot when the small tank was full and gasoline poured into the passenger's compartment if pumping was not stopped in time.

The asylum for theoreticians to which I was assigned when I arrived at Farnborough was called H department. I never knew what H stood for. It might have been Heaven, or it might have been Hell, so varied were the ideas of other departments of what went on there. It tackled problems wished on it by the War Office which the engineers did not care about. I will describe one job which was assigned to me and Melvill Jones (who was later to become the first Professor of Aeronautics at Cambridge) because it gives an idea of the kind of thinking that prevailed among those who were directing military operations at that time.

The French had been dropping steel darts called 'flechettes' on troops from the air. It was found that these whirled when dropped even though they would point into the wind when suspended at their centre of gravity, and we were asked to design one which would fall straight. We pointed out that a bomb which would project its fragments horizontally would have a much better chance of hitting standing men than darts dropped vertically, but it seemed a matter of prestige that we should have our own dart, and that it should be better than the French dart, so we designed one. After we had done this we took a few hundred of them in a satchel up a steel ladder inside a tall disused factory chimney so that we could watch them from above as they fell. They flew quite straight and after we had finished the experiment we climbed down the ladder. I went first, then Jones, then a man called Clark who carried the satchel containing the remaining darts. Suddenly when I was on the bottom rung and Jones was about 25 feet up we heard a whirring noise. We guessed that the darts had got loose and were coming our way. I jumped for the passage along which the burnt gases had come when the chimney was in use. Jones was very quick. He nipped round the ladder and jammed himself between it and the brick wall of the chimney, crossing his arms over his head to protect it. Meanwhile Clark was unconscious of what was going on. He had just started down the ladder when the bottom of the satchel had caught in the top rung and up-ended itself as he climbed down. No one was hurt.

Having designed the dart we were asked whether its terminal velocity was great enough to damage troops. The incident in the chimney might have supplied the answer but fortunately for us it had not. So we bored out a rifle and shot the darts with reduced charge at a leg of mutton hanging as a ballistic pendulum. By observing how far it swung when the dart hit it we could calculate the velocity of impact of the dart and observe its penetration at that speed. Finally we were asked how the darts would spread if a bundle of them were thrown from a plane. To answer this we got a pilot to throw a few hundred of them over a field from a few hundred feet. When this had been done Melvill Jones and I went over the field and pushed a square of paper over every dart we could find sticking up out of the ground. When we had gone over the field in this way and were looking at the distribution so revealed, a cavalry officer came up on his horse and asked us what we were doing. We explained that the darts had been dropped from a plane. He looked at them and, seeing a dart piercing every sheet of paper, remarked 'I should never have believed it was possible to make such good shooting from an airplane.'

Having finished that job the darts were never used; not, we were told, for the reason we had originally put forward against them, that they would be inefficient, but because they were regarded as inhuman weapons which could not be used by gentlemen.

After being employed for a short time designing instruments for use in the air I felt that I might be able to do a better job if I were taught to fly myself. When I put this point of view to O'Gorman he agreed with me but the army refused to sanction the proposal that a scientist should be taught to fly. Their comment was that I only had to ask a pilot and he would tell me anything I needed to know about the use of instruments in the air. The fact was that the pilots often did not understand the physical causes of the sensations they experienced while flying and told, as facts, their interpretations of the causes of their sensations. O'Gorman however was clever at getting his way by roundabout methods. He allowed me to resign from his Factory so that I would be free to apply to join the Royal Flying Corps as an uncommitted citizen, and said he would apply for me to return to Farnborough when I had passed out of the flying school.

In May 1915 I went to the school which used the space in the middle of the old Brooklands motor racing track as an airfield. I started to learn on the 1913 model Maurice Farman biplane [figure 6.1], then called a Longhorn or Rumpety. This, though it looks so old fashioned, was a good machine for beginners. Its maximum speed was about 60 miles per hour and it was said to stall at about 37. We aimed to land it at 42. Though its engine fairly frequently gave out in the air, its low landing speed made it possible to come down safely in many grass fields and if it did crash the structure in front carrying the elevator could act as a shock absorber. The wings had a wooden framework and were covered with fabric. They were held together with round-section steel piano wires and so many were needed that the mechanics used to say that when they had erected a machine they released a bird between the wings; and if it got out a wire was missing. My instructor was a flight sergeant who had been injured in a crash caused by one of his pupils. He decided that this must not happen again so if any of them exerted any appreciable pressure on the joystick, or dual control column, he reported that they were heavy-handed and would never make good pilots. Since very few machines were then available for training and many people wanted to join the air service, the threat that one would be turned out was to be avoided at all costs and many of this man's pupils took control over their machines for the first time on their first solo flight. The resulting manoeuvres were

Figure 6.1 *While at the Royal Aircraft Factory at Farnborough in 1914–17 Taylor learned to fly on a 1913-model Maurice Farman pusher biplane like that shown here. Later he wrote 'I am glad to have lived during the time they (aeroplanes) were simple enough for a reasonably intelligent amateur to appreciate them as mechanisms.'*

sometimes spectacular. I was more lucky than some, for the months I had spent at Farnborough had made me understand at any rate the theory of how the controls work, but my first solo flight ended in the sewage farm which you can still see alongside the Southern railway track near Byfleet. Though I was never much good as a pilot the fact that nothing very dreadful happened to me may have influenced the authorities to withdraw their objection to training scientists to fly; at any rate a little more than a year later several more from H department were trained, including Keith Lucas, William Farren, George Thomson, Lindemann and Melvill Jones.

The training course at Brooklands provided a restful interlude in an otherwise restless world. We would go early onto the field and if there was more wind than would put a match out flying was cancelled and we would go home, though we had to keep on the alert in case the weather improved. On one such occasion when I had risked going to my parents' home in St. Johns Wood for the night I was wakened by gunfire. I saw the whole sky lit up and realised that something spectacular was happening on the other side of the house. I ran across a passage and saw an airship which I thought must be a Zeppelin but

which turned out to be a Schutte Lanz, coming down vertically
in flames. I watched it till it came down behind a chimney of a
house 150 yards away whose garden backed on to ours. I noticed
that the ship appeared just the same length as the chimney pot.
When it became light enough to see I got out my sextant and a
compass and found that the chimney subtended about 35
minutes of arc and its compass bearing was NNE. I knew that
the German rigid airships were about 600 feet long and 600 feet
subtends 35 minutes of arc at about 11 miles. I laid off 11 miles
NNE from our house on a map and found that the nearest
railway station to the position so determined was on the Great
Northern line at a place called Cuffley. I bicycled to Kings Cross
and caught the earliest workman's train to Cuffley. On leaving
Cuffley station I saw several people walking in my direction so I
followed them and found the remains of the airship concentrated
in a space about equal to its cross section. It was a grizly sight.
The bodies of the crew had been covered but not yet removed.

As part of our training we made cross-country flights to give
us practice in finding our position on a map. Aerial navigation
was then very primitive but kindly stationmasters were very
helpful in many cases and set out the names of their stations in
large letters with whitened stones on their flower beds, so we
used to come down low in the hope of seeing such a station when
we were lost.

For the first few navigation flights we were passengers with a
pilot who navigated, and on one of these the engine cut out when
we were nearly over Basingstoke and we landed in a field. We got
out and I saw several children about. I asked them what they were
doing and they answered that they had come to see the aeroplane
land. Asked how they knew it would land there they said that
aeroplanes always did. Then some hospitable people came out
from a large house nearby and asked us in to lunch. They were
friends of my instructor. After a very pleasant lunch we found
that the engine had miraculously mended itself. I started the
engine by pulling on a propeller blade and we went on our way.

When I had graduated as a pilot in July 1915 I was sent on a
course described as 'artillery spotting'. One was instructed to fly
one's machine near a certain farm which was marked on a map and
watch for smoke puffs which would be let off by people on the
ground. One had to tap a key screwed to the cockpit and give in
Morse Code one's estimate of the distance and bearing of the puff
from the target farm. Wireless telephony had not then been
invented and a trailing wire aerial had to be released in the air

before one could send a message. I failed at my first attempt because I forgot to wind up the aerial and dragged it off in a hedge at the perimeter of the field while approaching to land.

At that time many of the pilots and instructors were more accustomed to handling horses than machinery and distrusted instruments; indeed some instructors went to the length of disconnecting the instruments, telling their pupils they should learn by the feel of the controls. This advice was particularly dangerous when near the ground because people tended to look at the ground to see whether they were side slipping. If for instance one took off against a wind and looked at the ground while turning off that course, observation of the ground would make one believe one was side slipping inwards if one was correctly banked. There was therefore a strong incentive to move the controls so that one would appear from ground observation to be flying without side slip. This meant of course one was really side slipping in the air and this might lead to a dangerous situation. This condition did not arise if one used instruments, but when machines were normally flying at speeds comparable with the wind speed it was a trap for the unwary to judge whether one was side slipping by looking at the ground, and a good many people were killed that way. Pupils were therefore warned against making turns near the ground. I had complete faith in my instruments and never looked at the ground till it became necessary to land on it. A rumour even went round that on one occasion I was practising turns, thinking I was at 2000 feet, when I saw a tree nearby and found that I had been watching the engine revolution counter thinking it was the altimeter.

The slow speed and light loading of aircraft in those days made it much easier to do aerobatic manoeuvres than it is with modern heavily loaded machines. The simplest manoeuvre was looping; one dived till one had achieved a good speed, then pulled the stick back and held it there till one was upside down when one should be nearly weightless. One went on holding the stick back on the downward part of the loop till one was flying level again. I was not a good pilot and that was the only acrobatic manoeuvre I ever tried. The first time I did not get enough speed in the dive and hung on the straps at the top of the loop, but my later efforts were better.

In the early days of the 1914–18 war the engineers had no confidence in the theorists because mathematical predictions about the forces exerted by air on moving bodies never seemed to give experimentally verifiable results. Experiments with models in wind tunnels provided the only data available to designers, and though models could be made *geometrically* similar to full-scale planes, model experiments could not be carried out under conditions of *dynamical* similarity. For this reason it was important to find something that could be measured both on a model in a wind tunnel and also on a full-scale airplane in flight. The overall weight of the airplane was known and this could be compared with the lift measured on a model in a wind tunnel, but the same method could not be applied to obtain the drag, or air resistance, because nothing was known about the efficiency of the airscrew.

It seemed that the simplest full-scale measurement that could be compared with model experiments was the distribution of pressure over one section of a wing, and on my return to Farnborough after passing out of the flying school I took on this work. I got the workshops at Farnborough to construct a wing for the BE2C which had been allotted to me so that a section a few inches wide was rigid. The rest was fabric covered. In this rigid section there were 20 small holes. Each hole was connected to the top of a glass tube mounted in a manometer in the cockpit, by means of a copper tube which passed through the wing. A manometer is a device for measuring air pressure by balancing it against the weight of liquid in a glass tube, and my manometer contained 24 such tubes. Since I was to make the measurements while piloting the airplane it was lucky I had a stable machine because I had to fly the plane for some time at a constant speed before making a measurement. The measurement was made by taking a photograph of all the tubes at once. The tubes were in a light-tight box which contained a pea lamp which I could light up by pressing a button when I had kept the speed steady for sufficient length of time to be sure the apparatus had reached a steady state. The pea lamp cast images of the alcohol in the manometer tubes on to sensitive paper which I wound on after each exposure. I took measurements over the whole range of speeds at which the aircraft was capable of flying level, namely 50 to 90 miles an hour. When the results were compared with wind-tunnel measurements made at the National Physical Laboratory the agreement was not good, but it was found that the N.P.L. had not made a proper allowance for the effect of the tunnel walls on their measurements. When they did, the two sets of results were very close. This experiment vindicated the use of models to obtain design data. I think that when my results were published (1916f) they may have been the first full-scale pressure distribution measurements to be recorded though many have been made since.

One of the problems which was tackled at Farnborough in those days was the construction of a theory of airscrews. Their sections were airfoils and very soon it was found that if the wind tunnel data for each section of the blade were applied as though it were advancing into still air, the calculated result did not agree with measurements made on a test bench. On the other hand the results could be correlated if it were assumed that the air approaching the airscrew was already in rotation before it reached the plane of the screw. Thus it was common practice to introduce a fictitious 'rotational inflow factor' which made the calculations agree with observation. This shocked me because Lord Kelvin had already shown that circulation cannot be produced by the pressure variations such as might arise owing to the motion of the blades and I wrote a note pointing this out. Shortly afterwards H. Glauert cleared up the difficulty and showed that though the circulation in the approaching ring of air was in fact zero the direction of the air approaching each blade was changed owing to the pressure field produced by all the other blades. Thus the air in the field close in front of each blade deviates in one direction and the air approaching the spaces between the blades is deviated in the opposite direction leaving the total circulation in rings of air approaching the airscrew zero.

This example is interesting because it illustrates a condition under which correct conclusions about aeronautical situations can be based on the classical theory of non-viscous fluids. The forces on bodies moving steadily through the air are due to the viscosity of air and the mechanism of the process is described by the viscous boundary layer theory, but when a fluid is moved only by pressure gradients without the intervention of viscous effects, circulation (that is, the integral of the tangential component of velocity round any circuit of fluid particles) cannot be produced. That is why I said the rotational inflow theory of airscrew blades is fictitious.

Though much of my work at Farnborough was carried out with the stable BE2C which was controlled laterally by means of ailerons, I did fly two machines which had no ailerons and were controlled by warping the whole wing. Wings in those days were not the rigid constructions one sees now. Some had no ailerons and were controlled by wires from the control column or joy stick which moved the longitudinal wing spars so that when the stick moved to the right the starboard wing warped or twisted to decrease the angle of its incidence and the port wing to increase it.

This control worked satisfactorily when the air was calm but it was very uncomfortable when the air was bumpy. The stick then sometimes waved about uncontrollably, nevertheless a small steady force on the stick would make the machine roll so the machine was quite controllable in spite of the discomfort involved. The reason for this behaviour was that the stick had to control the incidence of the whole wing, not that of the much smaller ailerons. I mention this because there are not many people about who flew warp-wing machines. George Thomson tells me that he flew one of the two that I flew but did not experience the discomfort of bumpy weather.

In the 1914–1918 war one of the great difficulties in flying was that as soon as one got into a cloud and could no longer see a horizon one had no means of appreciating the vertical direction and pilots were liable to come out of clouds in steep banks thinking they were still in horizontal flight. Some instrument was needed to tell whether one was turning because it was the acceleration in a turn which falsified one's appreciation of the direction of gravity. Gyroscopes had not yet been designed for use in airplanes and some instrument was needed to tell pilots whether they were turning. One was invented by Horace Darwin, a son of the famous biologist Charles. He had noticed that it would be useless to fit Pitot tubes to register the difference of speed at the two wing tips when turning, because the pressure changes in the pipes connecting them with a pressure measuring instrument in the cockpit would exactly neutralise the pressure difference in the Pitot tubes. On the other hand, if static pressure tubes measuring the actual pressure in the air at the wing tips were substituted for Pitot tubes, the centrifugal force in the connecting tubes would tell the pilot whether he was turning. Darwin's turn indicator was constructed on that principle. This accelerometer made it possible to fly safely through clouds. Of course people had flown through clouds before turn indicators were invented but few could fly continuously without the aid of a visual horizon. So well was this difficulty publicised that it became necessary to set up a committee, one of whose principal duties was to turn down inventions of artificial horizons. The members of this committee have been described as inverted Micawbers, alluding of course to Mr. Micawber who spent his life waiting for something to turn up. I remember one inventor who explained that he quite understood that accelerations of the airplane were indistinguishable from gravity, but his invention got over that difficulty. He explained that it was a tube which emitted smoke.

The smoke particles are free from the airplane as soon as they have been emitted, and by the well known property of cigarette smoke they therefore rise in a vertical line.

The ability to fly at night had not made it possible to do night bombing except on clear nights when wind corrections for navigation could be made using ground observations, and on such nights people did not like flying over enemy territory. In 1918 Bertram Hopkinson invented a simple method for night navigation in raids on objectives at small distances over which the wind could be taken as constant. This consisted in shining two searchlights vertically from points a mile apart on the line to the target. On most nights there is enough haze to show up the vertical beams and the pilot manoeuvred his plane into the line of the two beams and found the compass bearing which kept them in line as he approached. He flew on this bearing through both beams. On passing over the first beam he started a stop watch which he read on passing the second beam. Then knowing the ratio of the distance of the objective to the distance between the beams, he multiplied this ratio by the time of passage between the beams. Then he was to continue his flight on the same course for this calculated time, drop his bombs and go home. The gear for bombing Coblenz by this method was set up near Colombé, but peace broke out on the day before the operation was to take place so the method was never used.

While describing my aeronautical experience during the first World War, I should perhaps describe a balloon trip. The rather primitive anti-aircraft guns round London were often manned by very inexperienced crews and the military authorities needed some object in the air on to which they could direct their search lights, so they organised flights of free balloons at night. On one of these I was a passenger with the pilot and two crew members. We went up from Kennington Oval, a well known cricket ground in South London. Very shortly we got into clouds at 4000 feet, and since we were then useless to the searchlight crews our pilot released some gas so that we fell out of the cloud. Putting my head out of the basket I felt a strong upward wind and soon noticed the partially blacked out street lamps getting further apart. The pilot said we were going down too fast so he released some sand ballast. At first the sand grains fell more slowly than we did and would have appeared to us to go upwards if it were not too dark to see. After a short time however our downward motion stopped and the sand caught us up. It was like being in a sand storm. Though we sounded a powerful Klaxon horn to attract attention, the searchlights did not find us, and after about an hour and a half we realised that we had drifted out of the range of the searchlights. The pilot then let out some gas and lowered a rather heavy long rope which reached the ground and trailed along, keeping us at about 100 feet up. He shone a torch downwards and we saw the illuminated spot on the ground. Every time we passed over a tree or house the spot jumped upwards. After a time the light spot stayed down and the pilot moved it about to be sure we were over open ground. When he had satisfied himself that we were he released a good deal of gas and we hit the ground with a bump for which we had prepared by bending our knees. One of the crew got out and went to find out where we were and to use a telephone to ask for a lorry to come and fetch us and the balloon. He did not return for some time, so the second crew member got out to look for him and the pilot let out a little gas so that we rested very lightly on the ground. The pilot and I then sat in the basket talking in the dark till we heard a shout, and he shone his torch round to see who was there. The torch beam went out into space without striking anything and we found we had risen about 50 feet without noticing it. I suppose the surrounding air had cooled a bit faster than the balloon. This method of landing by trailing a rope could hardly be used now; there are too many power wires.

At the end of the 1914–18 war the Daily Mail offered a prize of £10,000 for the first direct flight across the Atlantic. Four firms entered machines which they had designed during the war, adding extra fuel tanks for the competition. I expected to return to Cambridge in October 1919 as a lecturer in mathematics, and had half a year in which to spend the gratuity handed out to demobilised officers by the Government, so I was glad when I was invited to join the Handley Page group, partly as a meteorologist and partly to teach celestial navigation to the navigator, a Norwegian named Tryggve Gran. It was not practical to set out for Newfoundland till the worst of the winter weather was over since there were no airfields and the ground would be very soggy. The four competing firms, namely Handley Page, Martinsyde, Sopwith and Vickers, sent out their expeditions by sea about the same time. Though I suppose they each expected to derive some commercial advantage, the competition was essentially a sporting event. We each had to make our own airfield and erect our machines in the open. As soon as we landed from the ships which had taken our expeditions to Newfoundland we each chose a site to make into an airfield. Vickers chose a site on high ground behind St. John's, and Handley Page chose a field at Harbour Grace about 60 miles from St. John's. Martinsyde went for a place called Quidi Vidi Cove just north of St. John's. Till we moved to Harbour Grace all the expeditions lived in the one decent hotel in St. John's, and every morning we went off early to the chosen sites of our airfields and worked furiously to get them ready before our rivals.

Shortly after we arrived all the competitors decided to have a party. This could only be done early in the month because of the peculiarities of the Newfoundland liquor law. When the government had introduced prohibition they had bowed to the protests of the doctors who claimed that alcohol was in some cases a useful medicine. The doctors were therefore allowed to prescribe alcohol for strictly medical purposes. In the first month of this arrangement I was told that one of them prescribed 30,000 bottles. The authorities then altered the law so that each doctor was only allowed to prescribe a limited number each month. The doctors used to sell their scrips, as they were called, for a dollar a time and their ration of scrips used to run out early in the month, so you had to get in early if you wanted to have a party.

The first competitor to try his luck was the Martinsyde outfit. They had loaded their machine too heavily I think; at any rate, it crashed before becoming airborne. The navigator, a man called Morgan, was injured, and when this news got around would-be navigators rushed up to try for his place from all over the world. One I remember came from Argentina. The next to try was the Sopwith with Hawker as pilot and McKenzie Grieve as navigator. I saw them take off and we heard no news of them for nearly two weeks, when they turned up in Norway on a freighter. Pious people remarked, 'How wonderful are the ways of Providence that arranged that a ship should be near in a deserted part of the ocean when their flight ended in the drink.' The fact was of course that when an oil pipe fractured they knew they could not keep in the air long and searched for a ship and came down alongside when they found one. It was a Scandinavian freighter which had no radio transmitter so no one knew about the rescue till she reached her destination.

The third to be ready was the Vickers with Alcock as pilot and Whitten Brown as navigator. When the Vickers took off I was close to the plane at the start of its run. It ran the full length of the field and disappeared below the level of the high ground of the airfield. Everyone near me thought it must have crashed but in a few minutes it came back over us at a height of a hundred feet or so and headed out to sea. As everyone knows, they crash-landed in a bog in Ireland. Brown was not well known as a navigator, indeed he pretended to be quite ignorant of navigation though I think that was a pose.

The Handley Page machine [figure 6.2] being bigger than the others took longer to erect. It was just doing its test flights when Alcock and Brown got across, so it very wisely turned south for New York. I was thus free to amuse myself, and travelled about in the United States. After a most exciting voyage in a canoe down the Snake River I boarded a train to Salt Lake City. I got into conversation with the headmaster of a Mormon school who asked me whether I had a job. When told that I did not he asked me whether I would go and teach in his school in Salt Lake City. I enquired how he came to believe I could teach anything, and he replied 'That doesn't matter a bit, all you have to do is to speak to my boys with an English accent.' However I had other plans.

Figure 6.2 *In 1919 Taylor acted as meteorological and navigation adviser to the Handley Page group competing for the* Daily Mail *prize to be awarded for the first non-stop flight across the Atlantic. The large Handley Page machine shown here was assembled in Newfoundland together with three other competitors, but was not ready to fly before Alcock and Brown succeeded in crossing in a Vickers aircraft.*

I have tried to convey some impression of what it was like being concerned as a scientist with aeronautics when flying was more of a sport than a profession. I have said nothing about the ideas I then had, and still have, about its effect on our life. In this connection may I quote a sentence from my Wilbur Wright lecture on 'Scientific methods in aeronautics' (1921d) given 50 years ago to the Royal Aeronautical Society. Describing some of the early scientists whose work contributed to aeronautics I wrote, 'They may even have been people who believed that rapid transport is on the whole a bad thing; that every increase in the facilities for travel merely decreases the size of the world instead of increasing one's range'. With these reactionary sentiments I still find myself in agreement, but I did not expect to find sympathisers in the Royal Aeronautical Society.

It is now the fashion to blame scientists for social evils which may arise from the use which others make of their work. I must confess to the old-fashioned belief that it is the business of our government and not that of scientists to control that use.

CHAPTER 7

Return to Cambridge after World War I

Life with Rutherford

I n October 1919 Taylor, or 'G.I.' as he became known to his friends, was appointed to a fellowship and lectureship in mathematics at Trinity College, Cambridge, after having been employed at the Meteorological Office in the immediate post-war period. A new phase of his life then began. As a bachelor Fellow, G.I. lived in the College, and had many friends among the Fellows there. Earlier in the same year he was elected to Fellowship of the Royal Society, in recognition of his significant contributions – not yet numerous, but all innovative – to meteorology and oceanography and aerodynamics. 1919 was also the year in which Sir Ernest Rutherford came to Cambridge as the Cavendish Professor of Physics. Rutherford was also a Fellow of Trinity, and he and G.I. became close friends. The flavour of College life around the dominating figure of Rutherford is amusingly conveyed by the following extract from G.I.'s Hitchcock Lecture at the University of California entitled 'A scientist remembers' (paper 1952d, unpublished):

> We were a polite society and I expected to lead a quiet life teaching mechanics and listening to my senior colleagues gently but obliquely poking fun at one another. This dream of somnolent peace vanished very quickly when Rutherford came to Cambridge. Rutherford was the only person I have met who immediately impressed me as a great man. He was a big man and he made a big noise and he seemed to enjoy every minute of his life. I remember that when transatlantic broad-casting first came in, Rutherford told us at dinner in Hall how he had spoken into a microphone to America and had been heard all over the continent. One of the bolder of our Fellows said 'Surely you did not need to use apparatus for that'.
>
> The thought of Rutherford coming to Cambridge calls to mind a bull in a china shop, but that simile is very wide of the

79

Figure 7.1 *Geoffrey Taylor at age 33, in 1919.*

mark. The china figures soon found they were in greater danger of falling off their perches through laughing with Rutherford than of being trampled by him. Though Rutherford was not much interested in mechanics he seemed pleased to provide space in the Cavendish Laboratory for a mathematician who was willing to try out experimentally results obtained by analytical reasoning. He gave me a room near his own, in fact he had to pass through mine to get to his, and I became, I think, a side show on to which he could divert a small part of the flood of visitors his fame brought to the Laboratory. Many of these were foreigners, but fortunately most of them spoke some English for Rutherford was not a good linguist. I only remember hearing him speak one sentence in a foreign language. He was showing his apparatus to a distinguished French mathematician who could speak no English. I saw him point to a certain spot and say 'ici les α-particles'.

Every Sunday morning Rutherford, F.W. Aston, R.W. Fowler and I used to play golf and often other friends joined us. Though Fowler was a scratch player and Aston fairly good, Rutherford was not good and I was just awful. Had we been serious golfers such a party would have broken up in a short time, but it met nearly every Sunday in term time from 1919 till Rutherford died in 1937. The fact is we really went to hear Rutherford talk, but the other golfers on the links did not have that advantage so we were sometimes unpopular. I remember one Sunday when Aston had just got results with his first mass-spectrograph he spent the first 5 holes discussing with Rutherford what the new atoms he had discovered should be called. By the time we had got to the 6th green it was evident that the party behind us had become really impatient, so Rutherford and Aston made a snap decision and adopted the name 'isotopes' which I think had been previously suggested by Soddy, and we hurried on to the 7th tee.

Rutherford seemed to take as much pleasure in detecting the causes of human reactions as those of nuclei. I remember once listening to him at a private party releasing whole bagfuls of cats and telling a story about X, a high official, the publication of which one would think must lead to X's suicide and the resignation of a Cabinet Minister. After a time he looked slightly conscience-stricken and said 'I almost let something out that time, but I like old X, he's very human'.

Rutherford was the most direct person I have ever known. When he said anything, one knew he meant exactly what he said.

Figure 7.2 *G.I. and Ernest Rutherford, date unknown.*

His forthright speech sometimes led him into what other men would have regarded as awkwardnesses, but his directness usually got him out of these situations. I remember once when he was dining in Hall he said to his neighbour in his loud voice 'You know these senior wranglers[1] never seem to amount to anything much in later life'. There was an awkward feeling among his neighbours at dinner because we all knew that two of those within hearing had been senior wranglers. Rutherford felt the slight tension. He leaned across to the astronomer Arthur Eddington who was opposite him and said 'You weren't a senior wrangler, were you?' Eddington had to confess that he was. Rutherford replied 'Well, Eddington, all I can say is that I am surprised you are such a thundering good chap'.

Rutherford inspired devotion in everyone who had the luck to work with him, and in no one more than in Peter Kapitza, who came to Cambridge from Russia about 1922. At first Kapitza did not understand Rutherford. I suppose in Russia it is inconceivable that anyone can really mean what he says, and Kapitza used to come to me with ingenious alternative speculations about what the Crocodile (Kapitza's private name for Rutherford) really meant by something he had said. I used to tell him that Rutherford meant exactly what he said and finally after some time Kapitza came to understand that this was always the case.

Back to research in fluid mechanics

G.I. produced a number of papers of significance during the first few years after his return to Cambridge. It was as if a coiled spring had suddenly been released. We can identify three separate major developments which deserve a brief description. The three themes all belong within fluid mechanics and are tidal oscillations, motion of solid bodies in rotating fluid, and stability of flow between rotating cylinders. We take them in turn.

The new work concerning energy dissipation in tidal oscillations stemmed from an application to oceanography of earlier work on the frictional force exerted on the ground by a turbulent wind. G.I. had found, from an analysis of the observed approach of the wind velocity to the gradient value at some height (direct measurement of the stress on a rough surface being very difficult), that the

1. See chapter 3

frictional force on unit area of the ground was about $0.002\rho V^2$, where V is the mean wind speed near the surface (1916a). V is not a well defined quantity since the mean speed varies slowly with height, but it was common practice to accept it as the mean speed at one or two metres above a surface with only small scale roughness, such as the sea or a grassy plain. The coefficient 0.002 did not vary much with wind speed or with the nature of the rough surface, and G.I. noted that the same formula applied, with a coefficient differing by a factor two or less, to the friction due to turbulent flow through a pipe with a rough wall (in which case V represents the mean velocity just outside the roughness elements) and to the friction on a river bed. This kind of evidence of the common properties of turbulent flow in systems of quite different nature intrigued G.I. – no doubt one of the by-products of his early introduction to meteorology and aeronautics – and I think he did try to account theoretically for the numerical coefficient 0.002, but neither he nor anyone else has yet been able to do that.

The new application of this formula for the frictional force at the ground was put into G.I.'s mind by the appearance in 1917 of a paper about the dissipation of energy due to tidal currents in the Irish Sea in which the author assumed that the flow was laminar and the friction at the sea bed was due to molecular viscosity. This was obvious nonsense to G.I., and, being at home both with turbulent flow and, as a sailor and navigator, with tidal data, he made a new estimate of the rate of dissipation in two different ways. The first and direct way was simply to use the above formula for the bed friction and the observed tidal stream velocity. The second and much more difficult method was to calculate both the total influx of energy into the Irish Sea from the North and South Channels over a tidal cycle and the work done by lunar attraction on the water over a cycle, the sum being the energy lost by dissipation over a cycle. G.I. found roughly equal values by the two methods, the new estimate of the dissipation being about 200 times that found previously (1919a, 1920a); and in the following year his colleague Harold Jeffreys used G.I.'s direct method of estimating the dissipation in tidal currents from the above formula for bottom friction to show that the total amount of dissipation in all the shallow seas and places on the Earth's surface where tidal currents are appreciable (amounting to about 56 times the dissipation in the Irish Sea alone) is roughly what is needed to account for the observed gradual increase in the length of the lunar day (1920a). This was a nice spin-off from an investigation of the wind velocity over a grassy plain which I am sure delighted G.I., although later estimates of the total dissipation and of the secular acceleration do not show such good agreement.

It had been generally believed that tidal friction played only a very small part in tidal phenomena, but G.I. observed that this new large estimate of the rate of dissipation implied that the energy of the tidal wave which comes out of the Irish Sea is only about a quarter of that of the entering wave. He went on to show that various tidal phenomena in the region of the South Channel, through which most of the energy flux occurs, are simple consequences of the superposition of two tidal waves of different amplitudes moving in opposite directions (1919a). This interest in tidal streams in channels led to two further papers, one a lengthy analytical investigation of the way in which a tidal wave of 'Kelvin type' (that is, a wave in which the fluid velocity is unidirectional and the lateral pressure gradient is balanced by Coriolis forces) is reflected at the closed end of a channel of uniform breadth (1921a), and the other a simple – but novel – demonstration that the observed tidal ranges at various points on the coast of the Bristol Channel agree well with the values calculated for a tidal wave progressing up a channel with linear variations of both depth and breadth (1921c).

The second of these research themes in the immediate post-war years was the motion of bodies through fluid which is rotating as a whole, and was motivated by a wish to find some circumstances involving rigid boundaries in which the predictions of theoretical hydrodynamics would be confirmed by experiment, in contrast to the classical inviscid treatment of steady flow past a solid sphere or circular cylinder. G.I. had published one paper on this topic early in the war years (1917f), and he returned to it in several more just after the war (1921f, 1921g, 1922c, 1923f). The properties of rotating fluid systems later became a popular topic in the literature of geophysical fluid dynamics, and G.I.'s early work has been widely referred to, somewhat loosely, and rightly regarded as a foundation for future developments. The work possibly has been given more praise than, strictly speaking, it deserves, for in fact these papers do not do all that appears to be claimed, and the key theoretical result that slow steady motion with dominant Coriolis forces in fluid rotating as a whole is necessarily two-dimensional, in the sense that material straight lines parallel to the axis of rotation remain straight and parallel,[2] had been published a little earlier by J. Proudman.[3] The main new contribution of the papers is an extremely simple laboratory demonstration of a startling consequence of the Proudman theorem, namely, that when a body such as a sphere is moved slowly, in any direction, through a body of liquid which is rotating, the sphere takes with it the fluid in a cir-cumscribing circular cylinder with generators parallel to the axis of rotation so as to keep the motion two-dimensional in the above sense. G.I. derived his satisfaction from the interplay between applied mathematics and experiment, and he frequently alluded to this parti-cular experiment as one of the most satisfying, as well he might. The fluid in the circumscribing cylinder is usually called a 'Taylor column' nowadays in recognition of his early perception of the dramatic experimental consequences of dominant Coriolis force, but it would probably be more just to refer to it as a Taylor–Proudman column. The great red spot in the atmosphere of the planet Jupiter is thought by some astronomers to be a spectacular manifestation of such a column in rotating fluid with dominant Coriolis forces.

We come thirdly to G.I.'s grand and definitive investigation of the stability of steady flow between concentric circular cylinders in rela-tive rotation (1923c). The notion of stability of fluid flow systems at that time was imprecise, and, although the experiments of Reynolds and others had shown that under certain conditions some theoretical steady flow systems could not be realized and were replaced in practice by a permanently fluctuating turbulent flow, there was no case of steady flow for which a critical condition for stability had been both

2. This is equivalent to the more transparent statement that, as a consequence of Coriolis forces being strong, the divergence of the (vector) component of the fluid velocity in the plane normal to the axis of rotation is small everywhere.

3. *Proc. Roy. Soc. A* **92**, 1916, p. 408.

calculated and established from observation. G.I. sought always to clarify general principles by a study of particular cases, and he therefore looked for a type of steady flow field which might allow both these developments, wisely going outside the class of unidirectional flow systems which appeared to be considerably more difficult mathematically. I think he perceived intuitively that steady flow between concentric cylinders, for which an algebraic criterion for stability to inviscid small disturbances had recently been given by Rayleigh, would allow a normal-mode analysis of the behaviour of small disturbances, and that the interesting critical disturbance that neither grows nor decays would have the simple property of not propagating in the direction of the cylinder axes. He also anticipated correctly that the role of viscosity would not be singular in any way.

The normal-mode disturbance is sinusoidal with respect to axial distance and is a sum of Bessel functions of the radial coordinate, and although determination of the properties of the disturbance that is neutral at the smallest possible value of the angular speed of the inner cylinder is relatively straightforward it is nevertheless a long calculation. Taylor was able to carry it to the point of approximate numerical results for the case in which the width of the fluid annulus is relatively small. At his request, Tom Cherry, an Australian mathematician then a Fellow of Trinity,[4] checked all the calculations for him. Taylor planned experiments on the stability of this flow system with equal care, and recognized in particular that it is necessary to use cylinders whose length is very large compared with the annulus width. Determination of the critical conditions was made possible by the ingenious device of smearing the inner cylinder with dye (water being the working fluid between the cylinders) and making the outer cylinder transparent. The angular speed of the outer cylinder was held fixed, and that of the inner cylinder was raised slowly; and when it reached the critical value the dye was observed to be convected across the annulus at a number of positions regularly spaced along the axis, thereby giving both the critical speed of the inner cylinder and the wavelength of

4. And later, Professor of Mathematics at the University of Melbourne and my teacher there.

the critical disturbance, both being in excellent agreement with the calculated values (see figures 7.3, 7.4 and 7.5). Taylor explored a range of cylinder speeds, and found that for steady slightly supercritical conditions the growing disturbance ultimately settled down to a steady state, evidently as a consequence of nonlinear effects. He also observed the curious forms, not yet explained, that this stationary toroidal vortex disturbance took as the speed of the inner cylinder was raised further, before the motion finally became turbulent. It was a tour de force which, more than any other single paper, established hydrodynamic stability as a distinct field. G.I. sent an off-print of his published paper to W. McF. Orr in Dublin, who, like Reynolds, had been pursuing the alternative plan of calculating whether the energy of a disturbance of assumed form would decrease or increase, and Orr responded with 'I admire your recent paper greatly.'

There is an interesting application of G.I.'s results for the case in which the outer cylinder is stationary and the gap between the two cylinders is relatively small. The undisturbed velocity distribution here approximates to the boundary-layer flow over a concave rigid surface, and so may be expected to be unstable for sufficiently large

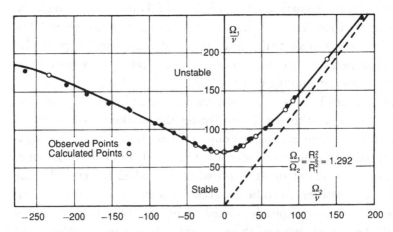

Figure 7.3 *Comparison between observed and calculated critical speeds for the case when* $R_1 = 3.55$ *cm and* $R_2 = 4.035$ *cm.* Ω_1 *and* Ω_2 *are the angular velocities of the inner and outer cylinder respectively. The broken line shows the stability criterion for inviscid fluid. (From 1923c)*

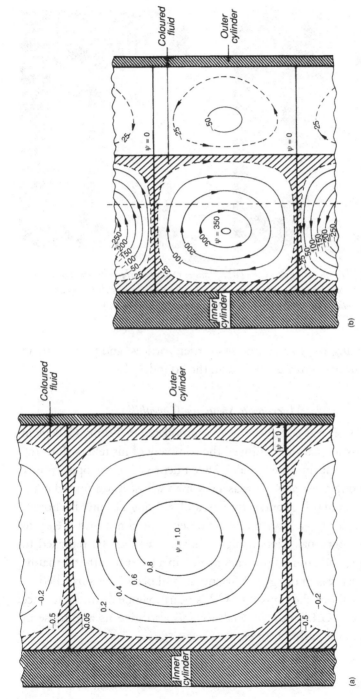

Figure 7.4 *Streamlines of the toroidal motion at a slightly super-critical speed of rotation of the inner cylinder. Case (a), cylinders rotating in the same sense; case (b), cylinders rotating in opposite senses. (From 1923c)*

(a) 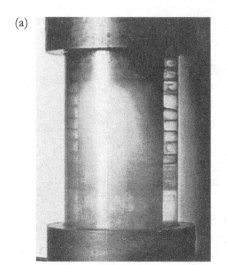 (b)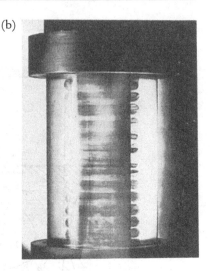

Figure 7.5 *The spreading dye. Case (a), cylinders rotating in the same sense; case (b), cylinders rotating in opposite senses. Note that the cells extend only part-way across the annulus in this latter case, as in figure 7.4(b). (From 1923c)*

speeds, giving rise to streamwise Görtler vortices[5] and perhaps to an early transition to turbulent flow in the boundary layer.

Marriage to Stephanie Ravenhill

Although G.I. was comfortably accommodated in Trinity College in these post-war years and enjoyed the company of his fellow dons (all of whom at that time were male), he felt deprived of the opportunity to meet women socially. He was a shy man whose mind turned naturally to material things and activities, and, although I think his boyish gentle manner must have pleased women, he did not find it easy to form close personal relationships. He was 33 when he returned to Cambridge, and after a few years he said in a letter to a friend that he felt his heart was in danger of drying up and becoming shrivelled. A few words about the revival of G.I.'s heart may be of interest, since they concern G.I.'s human qualities and also give us a glimpse of social life in the early twenties.

5. H. Görtler, *Nachr. Ges. Wiss. Göttingen, Math.-phys. Kl*, **1**, 1940, p. 1-26.

G.I. had in fact already become friendly with a Miss Mabel Lapthorn, a woman three years younger who studied art under his father and appears to have lived in London for most of her life. Sixteen letters written by her to G.I. in 1915 (when both of them lived at Farnborough) have survived, and, although they are mostly undated and difficult to decipher, it is clear that there were earlier letters. These letters are a little pretentious, and highly personal, and are full of argument by quotation; quite unlike what one imagines G.I. would have written. One senses her frustration at being unable to evoke a similar personal note from him. The sequence of available letters ends abruptly, almost certainly as a consequence of later letters being destroyed. In 1931 Mabel did a fine pencil drawing of G.I. and presented it to him at Christmas. It commonly hung in his study at 'Farmfield'. The extant part of the correspondence comes to life again in 1961, when she was still unmarried, and there are 63 letters from Mabel spread over the next 13 years, the last letter being written only 9 months before his death. On 9 February 1974 she wrote 'I remember with much pleasure the walks and outings on wheels that we used to have but I also cannot forget the endless arguments and here we are, just on 86 and 89, still at it! Let's part friends, and then somewhere and somehow meet again, still friends.' There is some resignation expressed in the letters from Mabel, and one wonders whether G.I. was simply being kind to her. In any event it shows remarkable patience and devotion on his part that he participated in a correspondence lasting about 60 years.

In early 1924, when G.I. was 38, he became a close friend of Ursula Nettleship, a singer whose sister Ida married the painter Augustus John. Miss Nettleship was evidently known to and accepted by G.I.'s family, and they had common friends in Owen and Mary Anne O'Malley[6]. It appears that G.I. declared his affection and regard for her, but she replied that she did not believe they were well suited to each other and hoped they would continue to be just good friends. This was a great disappointment to G.I., but nevertheless he adhered to plans which had been made in the spring

6. Who wrote several novels under the pseudonym Ann Bridge.

of 1924 for an ambitious summer sailing party in G.I.'s boat *Frolic* off the west coast of Scotland, consisting of himself as skipper and, at different times, Ursula, G.I.'s brother Julian and his mother Margaret, Edgar Adrian, George McKerrow (a sailing friend), the O'Malleys, and Carmelita Hinton (wife of G.I.'s cousin Sebastian Hinton) and her three small children from USA. The few surviving letters from Ursula to Geoffrey during the period June–November 1924 indicate that she was very concerned about hurting his feelings, that she was unsure whether he understood her sufficiently, and that the 'lack of rapport' between them during the cruise had caused some tension and mental discomfort for her. Ursula concludes a letter of reappraisal of their relationship written in September 1924 with the words 'You don't know how nice it is to know you're there, a real friend now and untake-away-able, and that I need never pretend anything to you any more.'

The loss of Ursula Nettleship was soon made good however. On 23 August 1924 G.I. began a correspondence which was clearly more than casual with Stephanie Ravenhill, a teacher of modern languages at King Edward's School for Girls in Birmingham whom he apparently met in Scotland while on the summer cruise. Stephanie, who was three years older than G.I., was less sophisticated than Ursula and her kindly no-nonsense manner appealed to him. Their love developed through the course of 76 letters from Geoffrey to Stephanie and a comparable number from Stephanie to Geoffrey (although fewer of these have survived), a skiing party composed of Stephanie and Geoffrey and a few of his friends at Arosa in December 1924, and several meetings in London or Cambridge or Birmingham, with either Stephanie's sister or her brother as chaperone.

Here is Geoffrey in the fifth of his letters to Stephanie, following up a suggestion that they might meet in London and go to a theatre. He quickly gets on to a discussion of Donne's poems which in his previous letter he says he has 'just been introduced to' and which give him the opportunity for the kind of innocent literary love-making by proxy that was common when social conventions were very restrictive.

Trinity College, Cambridge. 1 Nov. 1924

Dear Miss Ravenhill,

It will be jolly to see you on the 15th. Have you any idea as to what you would like to see then? Sob stuff, legs or high brow – those are the main divisions of the English stage. I think we might cut out the sob stuff. Would you care for the Sheridan play at Hammersmith or a Shaw? Or would you prefer George Robey who I must confess always amuses my somewhat vulgar mind. Ask your sister, whose tastes of course I cannot guess. I am in favour of the Sheridan if Edith Evans is in it.

I am glad you like Donne, you didn't say which part of him struck you most. I am interested in your 'him'. Do you regard books as people and books you like as friends? I know some people do. With Donne I also should have used the word 'him' but with no other poet that I know. What I mean is this. Other poets, greater poets than Donne – Shakespeare, Milton, Shelley for instance – give me pleasure. They show me scenes, characters, emotions, so that I seem to see exactly what the poet saw, but I have no idea at all of the poet himself, as a man. After reading most of Shakespeare's plays, and all his sonnets, I have only a vague idea of someone looking rather like Hall Caine. The artist and the philosopher have obliterated the man. Perhaps this is the highest form of art but I can't feel friendship for a philosopher or artist as such.

With Donne the case is exactly reversed. There is a real man talking to a real woman, or rather to a series of different ones. I can see Donne himself. If I met him in the street I should recognise him and should know what to say to him. If I were a woman I feel convinced I should fall in love with him on sight. Curiously enough though I get such a vivid impression of the man himself I get no idea at all of the characters of the women he is talking to. I only get the impression that they exist in the flesh and are not idealised or abstracted by the poet for the purpose of making verses.

I can't quite analyse what it is I like so much about him. Possibly it is the habit of mind which spiritualises the flesh and at the same time 'carnalises' (if there is such a word) the spirit. Well I suppose I shouldn't write such things to a 'school marm'. You might find yourself asking your little pupils to write essays on 'the place of bodily attraction in spiritual love' or some such title and what would the head mistress say then?

. . . .

Yours, Geoffrey Taylor

Geoffrey obviously contemplated marriage, but he hesitated to propose this in view of his disconcerting previous experience. He had believed his love for Ursula was deep and strong, but nevertheless it had not lasted. How could he be sure that his love for Stephanie would be more permanent? Geoffrey felt this dilemma was ultimately resolved by a meeting with Ursula, after which he wrote as follows to Stephanie on 21 May 1925:

My dear Stephanie,

Yesterday evening I saw Ursula for the first time since last summer. We dined together and had a long talk, but almost as soon as we met I knew that the thing I feared is dead and that I shall never want her again as I did before. We had a jolly evening together meeting on quite a different sort of plane. I found we had less constraint than before and in a curious sort of way more in common.

It makes me feel free and happy because it makes me feel sure, as never before, that the love I feel and have felt for you for some time is real and permanent. You know that on my side the thing which has kept us apart has been the fear that things which I now feel to be permanent may not really be so. I felt before that I should always want Ursula, just as (but in a different way) I now feel that I shall be miserable if I have to do without you. My dear I must put this down so as to be really honest with you. Actual knowledge of the future is impossible. Feelings and experience are the only guides and experience tells me that feelings however strong are not infallible.

I may change, you may change, you simply cannot know you won't, but it seems to me that there is, or may be, an enormous difference between now and before. If you can return the love which I have for you I believe that the 'feelings' will be right and the 'experience' inapplicable and that we should be ever so happy together. Now my dear you know the whole story. Will you marry me? It's a risk of course, but if when you search your own mind you really feel you care for me as I care for you, it's just the risk that is more worth taking than any other in the world.

. . . .

Yours, Geoffrey

Well, the logic of this letter may not stand up to analysis, but there is no doubting the honesty and decency of the writer. Unfortunately Stephanie's letters written between December 1924 and June 1925 are missing, but it clear from Geoffrey's next letter on 24 May 1925 that her response to this proposal from a distinguished scientist was an immediate and enthusiastic yes. Subsequent letters are taken up to a large extent with the selection of fittings and furnishings in the house in Huntingdon Road in which they were to live. They were married on 15 August 1925, at which time the correspondence ends, and immediately left for a sailing holiday in *Frolic* off the west coast of Scotland.

Stephanie (figure 7.6) was undoubtedly a very suitable partner for G.I., and their marriage was a happy one. She contributed a great deal to his well-being and thereby enabled him to use his scientific ability fully. There were no children. She knew little of G.I.'s academic and scientific world, but she was devoted to him and proud to be his wife. For a time Stephanie happily shared in his adventurous activities – sailing, skiing, and exploration of remote parts of the world – although there is no record of her leaving Britain for any such purpose after their tour in Borneo in 1929 (see chapter 10). She seldom accompanied G.I. to conferences but he always wrote cheery notes to keep in touch when he was away. Here is the first of numerous such letters in which he chats affectionately about the journey, the weather and the location, and says – invariably – that his 'paper seemed to go down well', this one being written from Zürich where G.I. was attending the Second International Congress of Applied Mechanics:

13 September 1926, Zürich

Dear S,

Bones[7] and I had a very comfortable journey. At Boulogne we found ourselves with 3 others belonging to a party of Christians. Soon after the the train left however a Christian young general (at least he wore a uniform almost identical with that of a brigadier) came in and removed the party to another carriage where he said he could make them more comfortable. He certainly made us more comfortable because we had the carriage to ourselves all the rest of the way. He was the most hearty man I have ever seen; every ten minutes or so he poked his head in and said that the ticket collector was coming along, or that the franc is now worth only $1\frac{1}{2}$ d, or that we must now line up for dinner if we want any, etc. At Zürich we took a taxi to the Technical University because, of course, Bones had forgotten the name of his host and had lost the letter about it. After trying 49 locked doors we at last found a 50th which had a bell push near it and we succeeded in rousing the porter. After that all went smoothly and I came to this house where I found an extremely pleasant host expecting

7. B. Melvill Jones, Professor of Aeronautical Engineering at Cambridge.

Figure 7.6 *Stephanie Taylor, née Ravenhill.*

me. You would like this house very much. It is lined throughout on the first floor with really beautiful dull-polished wood. The bedrooms are papered with light coloured paper and have ordinary white ceilings. I can reach my hand out of the window and pick ripe peaches growing on a tree up against the house. The really astonishing thing however is the extraordinary cleanness and tidiness. The moment one leaves a room, either a bedroom or dining room, the maids seem to go in and put everything straight. I got up yesterday to fetch a cigarette after breakfast and returned in 2 minutes thinking I would have a 2nd cup of coffee but when I got back the dining room looked as though such a vulgar thing as food had never sullied the chaste polish of the table – not a vestige of a crumb was to be seen. The street outside is so pure that I hardly dare throw a peach stone there so I am faced with the alternatives of eating them or packing them in a paper parcel and bringing them back to Cambridge.

I read my paper today and it seemed to come off all right. I was chairman in the morning and my paper was in the afternoon so I have been very busy all day. This evening we are dining with Professor Meissner and tomorrow with another Professor.

The English are very badly represented here because Southwell, Lamb, Bairstow and Griffith who ought to have been here have none of them come so Bones and I have to do our best to look like a large crowd.

Dr Fokker presents his compliments to you.

I think it is a good idea to have a separate holiday because I never knew how much I should miss the little creature. It seems lonely to wake up and have no little nose to watch as it wiggles among the pillows.

We haven't yet decided when I will go home.

G.

A few years after their marriage G.I. and Stephanie planned and had built a house in Huntingdon Road, 'Farmfield' (figure 7.7), in which they lived for the remainder of their lives, enjoying especially the large and well-stocked garden, where there was full scope for G.I.'s life-long interest in botany, and the distant views across the fields of the University farm. Stephanie was brisk, efficient and warm-hearted, and she looked after the domestic arrangements with the help of a resident house-keeper, Miss Gladys Davies, who remained with the Taylors for many years and cared for each of them in their later lives.

Figure 7.7 *Farmfield*

Stephanie was kind and hospitable in particular to the many people who came to Cambridge to meet and talk and work with G.I., although her firm views on etiquette were a little disconcerting for the younger people, especially those from Australia. I recall her regular Christmas parties at 'Farmfield' with punch and hot mince pies, very English in atmosphere, so we thought, and diverting for me as an impressionable research student from overseas since any one of the people with whom I was playing games might turn out to be a knight or an FRS or even both. She died in June 1967 after some years of mental incapacity. During that period G.I. was seldom away from Cambridge.

CHAPTER 8

Sailing

T he development of G.I.'s passion for sailing has already been noted. 'Passion' is the right word, for the handling and navigating of a yacht in strange waters and all weathers was the love of his life. Sailing is well adapted to the appreciation of nature, especially when it is close to land, and this too was an important part of the appeal of sailing for G.I.

Clear evidence of an early love of the sea appears in the delightful letter (figure 8.1) written by Geoffrey as a young boy to his father while on holiday at Seaview on the Isle of Wight. Note the precision of his descriptions of the events on what was undoubtedly a day of bliss.

G.I.'s career as a yachtsman began with the building of a $13\frac{1}{2}$ foot boat in his bedroom during his last year at school, in which he sailed up and down the Thames, sometimes sleeping on board overnight (see chapter 3). This was fun, but G.I. had more ambitious plans. The value of the Smith's Prize that he was awarded in 1910 by the University of Cambridge was £23, and by adding £32 from his savings he was able to buy a 5-ton yacht, *Elaine*. This was his first seagoing boat, and in it he sailed from Ipswich to Devon and back. On another occasion G.I. took on two classicist friends, W.C. Cleary and Walter Lamb, as crew and sailed down the Channel from Folkestone. However they ran into heavy weather near Eastbourne, and the two crew members became very sea-sick. Cleary was so ill that it was thought necessary to get him ashore, which they were able to do with the help of an old fisherman who rowed out to where *Elaine* was at anchor. The fisherman also advised them that if the wind changed a little they might not be able to maintain their position, so G.I. turned around and ran back to Folkestone at high speed where he and Lamb left the boat and returned to London by train.

99

Seaview

Teusday Isle of Wight

Dear Father

I have a lot to tell you about this time. On Saturday we went to Portsmouth to meet Mr Perrin. We went over Nelsons "Victory" which I expect yo have seen And we went over a modern "Warship "HMS Trafalgar" first class battleship, one of the biggest we have. I know you don't very much care for them, but they are very interesting. The sailors play very comic cricket. The ball for intance is made of brown paper stuffed with cotton-wool and the bat was half a packing-case lid. But I must get on to the yachting expedition. Well, yesterday we (Sydney, Maurice, Mr Perrin & I) went with Mr Walker the most ripping boatman I have ever known. Mr Walker is the Boatman who takes care of the "Waterwitch", Mr Nasmiths smallest yacht (he has got two yachts the "Cheta" a big one & the "Waterwitch" a little one). It was the "Waterwitch" in which we went to Bembridge yesterday. The wind was against us so we had to tack. We took one large tack 3 across the Solent or Spithead I forget which it is.

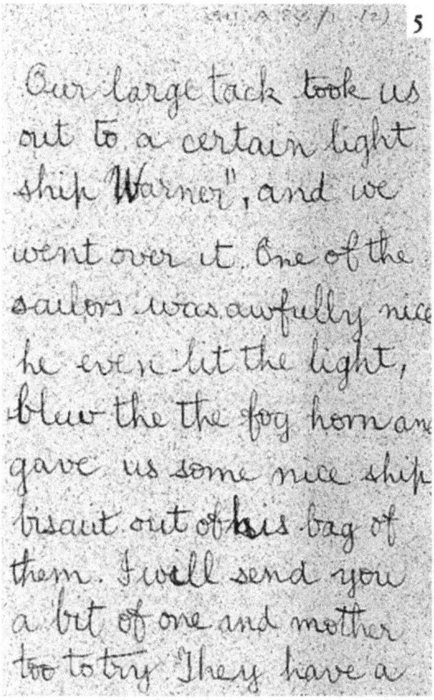

Our large tack took us out to a certain light ship "Warner", and we went over it. One of the sailors was awfully nice he even lit the light, blew the the fog horn and gave us some nice ship bisaut out of his bag of them. I will send you a bit of one and mother too to try They have a

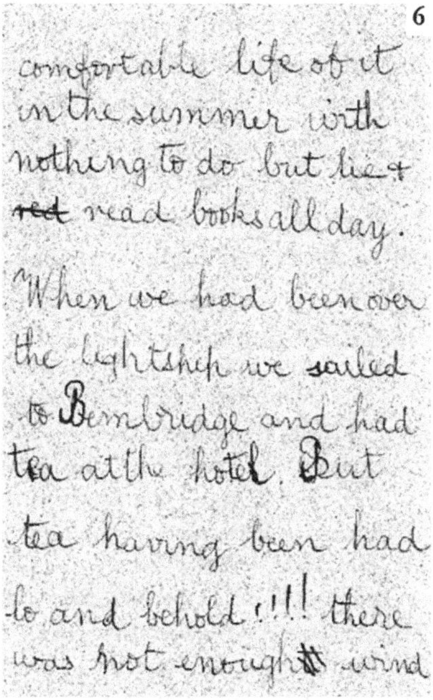

comfortable life of it in the summer with nothing to do but lie + read read books all day. When we had been over the lightship we sailed to Bembridge and had tea at the hotel. But tea having been had lo and behold !!!! there was not enough wind

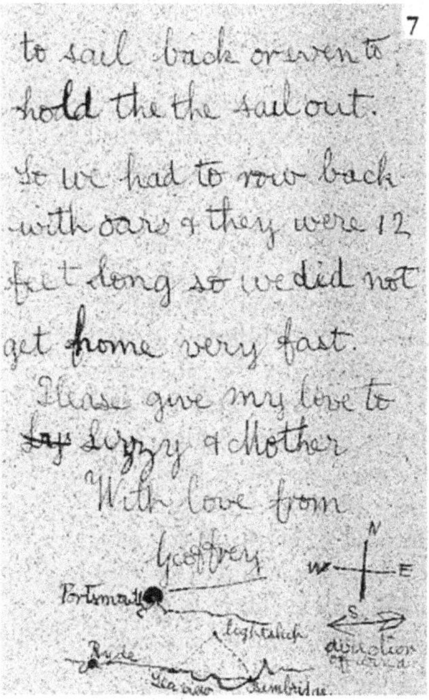

to sail back or even to hold the the sail out. So we had to row back with oars & they were 12 feet long so we did not get home very fast. Please give my love to Lizzy & Mother With love from Geoffrey

Figure 8.1 *A letter written by Geoffrey to his father while on holiday at 'Seaview' on the Isle of Wight.*

Coping with bad weather is part of the adventure for a sailor, but a
land-lubber might well feel that occasionally the elements come peri-
lously close to winning the battle. Consider for instance the fate of
the next and larger boat, a 17-ton cutter, *Seaflower*, bought by G.I.
near the end of 1912. On his return from the *Scotia* expedition in
August 1913 G.I. had *Seaflower* fitted out and began cruising on the
east coast. G.I. tells the following story of *Seaflower's* last cruise in
September 1913, making light of the fact that it ended with him and
a friend clinging to the wreckage of the boat for 14 hours.

> One weekend I had Cleary as my crew and started to sail from
> Pin Mill on the River Orwell to Walton Backwater. To enter that
> harbour one had to sail through a channel inside the Pye Sand
> and the channel was marked by three small buoys on the
> starboard and three on the port. There was a strong wind and a
> good deal of sea when we got to the channel, and we could only
> see a buoy when we were fairly close to it. We passed two buoys
> on the starboard and three on the port. Then we saw a sixth buoy
> right ahead and naturally assumed it was the third buoy on the
> starboard hand. Accordingly I altered course to pass to the left of
> it. We ran aground close to it, and since the wind and sea were
> driving us on we could not get off. It was 5.30 p.m. when we
> struck and nearly high water. As the wind and sea got up we
> bumped heavily on the sand but the boat held together. Since
> there was a channel between us and the shore we could not walk
> ashore when the tide left us, and had to stay near the yacht.
> Meanwhile the wind and sea rose and when the tide came in
> again in the night we had to climb onto the boat which began to
> disintegrate. We found afterwards that the whole of the
> starboard side parted and drifted away so that we were left on the
> wrecked port side which also had the mast. As the tide rose in the
> early morning we could feel the boat shifting on the sand and
> were afraid it would move into the channel. Fortunately at high
> water, when the side of the boat on which we were sitting was
> about at the mean water level, the wreck seemed to begin to
> move away from the channel and finally the tide receded and we
> climbed down onto the sand. At the time when the water was
> highest I was washed off the wreck but was clinging to the
> shrouds. My hands slid up the shroud and I was able to get back
> to the wreck, pulling myself along the wire, before the next big
> wave came.
> We were rescued later in the morning by a boat which came

out of the Walton Backwater. I could not understand how I had come to make this navigational error till a few days later after an account of the shipwreck had appeared in the London papers a man wrote that he had nearly done the same thing a week before. He explained that it was because the third buoy on the starboard hand had disappeared and that local people had placed a 4th buoy on the port hand to mark a place where a post had been run down by a barge some time before. This was the only occasion when I had a serious stranding.

Cruises around the British Isles and beyond

World War I put an end to sailing on the open sea, but in 1919 or 1920 G.I. bought his next boat, *Sorata*, which was an 11-ton yawl, with no motor. G.I. usually kept a log of his cruises, and hand-written logs for most of his cruises have survived. The first of his cruises on *Sorata* was in June 1920, and the log was written up as an article for publication in *Yachting Monthly* with the title 'Navigation notes on a passage from Burnham-on-Crouch to Oban'. The main purpose of the article was to point out that, contrary to common opinion among yachtsmen, sextant observations of the sun could be of great value. The second cruise on *Sorata*, in 1921, was a more ambitious journey from the Shetland Islands to Norway with one friend (probably George McKerrow) as crew. There is no surviving log of this cruise, but in this extract from his later entertaining article 'Amateur scientists' (1969d) G.I. shows how an amateur navigator was able to make an original contribution to that ancient science.

> As we came near to the coast of Norway, the sky was overcast, I could get no sights, and we approached the Skjaergaard in a thick fog. The Skjaergaard is a line of offshore rocks and islets some miles from the mainland of Norway and extending hundreds of miles along it. I had sailed nearly 200 miles across the North Sea when I saw, under the fog, white water washing on a rock about half a mile away. One can often see farther under a fog than through it. I could not see the rock itself and, of course, could not tell which one of the many thousands in the Skjaergaard it was. I knew by dead reckoning to within about thirty miles where my 200-mile course must have put me, but that is not good enough for penetrating the Skjaergaard. Nowadays a navigator would use bearings from a radio beacon

(a)

(b)

Figure 8.2 *(a) G.I. in one of his earliest boats, Sorata, sailing in a Norwegian fjord in 1921; (b) G.I. and Stephanie leaving the Lofoten Islands on 24 June 1927, in Frolic, the other member of the crew being George McKerrow.*

and readings from a depth sounder to locate his position, but no such navigational aids then existed. But the chart of the coast where my dead reckoning placed me showed the outer edge of the line of rocks to run almost exactly North and South, with none of the really dangerous underwater rocks outside that line. I could therefore safely steer due north. I did so, and noted the reading on my patent log (which records total distance sailed).

As we sailed along, the fog varied; sometimes we could see about a mile and a half and sometimes only a hundred yards. Now and again, when the fog lifted slightly, we caught a dim sight of a rock or of white water washing on one. Whenever that happened, we took a compass-bearing of it before it disappeared in the fog, and read the patent log to find how far we had moved along our northerly course. When we turned north, I got a sheet of tracing paper, ruled a line on it to represent our course, and laid it over the chart. Every time we got a bearing, we drew that bearing-line on the tracing paper from the point on our northward course where our log put us. We kept moving the tracing paper over the chart with the course line north-south and trying to find a position for which all the bearing-lines intersected some outer edge of rock. After a few hours, with the bearings of eight or nine rocks or water-washes drawn in, we found such a position, and saw that the rock then visible had a sea mark, or vaarde, on it, partly confirming our identification. The chart showed that this vaarde, if the right one, marked the entrance to a narrow channel among the rocks used by the local fishermen to reach the inner channel, called the Hjeltefjord, which runs northward from Bergen. We took this channel, and when we got through it we could not see the other side of the Hjeltefjord. It was a small triumph, but now, with radio beacons and depth sounders, the amateur navigator perhaps needs less ingenuity.

G.I.'s next boat, *Frolic* (see figure 8.3), purchased in 1923, was the one that gave him the most pleasure. *Frolic* was a 19-ton cutter, 49 feet overall length and 10$\frac{1}{2}$ feet beam, with an auxiliary engine and roller reef gear (as on the *Scotia* – see chapter 5). G.I. made many interesting and enjoyable cruises in *Frolic*, beginning with a shake-down short cruise up the east coast of England with George McKerrow and Peter Kapitza as crew. The next one, during the summer of 1924, involved the transport of many of his friends and members of his family between different points on the west coast of north Scotland and has already been mentioned in chapter 7. G.I. prepared a very readable article about this cruise entitled 'Extracts from the log of *Frolic*, 1924' which was published in the *Royal Cruising Club Journal* (1924b). There also exists an appreciative complementary account of the cruise in typescript form by one of the passengers,

Figure 8.3 *Frolic, under sail, in which G.I. made many long cruises in the twenties and thirties.*

Carmelita Hinton, wife of a son of George Boole's eldest daughter and a resident of the USA. It seems that while G.I. was on the island of Skye during this cruise he met Stephanie Ravenhill, as has been related, and when he and Stephanie were married in August 1925 he could think of no better way of spending their honeymoon than to introduce Stephanie to sailing in these same waters.

G.I. also was keen to repeat his earlier cruise in *Sorata* along the very rocky and rugged coast of Norway. During the summer of 1926 he sailed along the southern coast of Norway and visited a number of places there. Stephanie would undoubtedly have been with him, and presumably at least one other person, but no records of the cruise seem to exist. We do know that he finally left *Frolic* at Kristiansund for the winter, with the intention of returning in the following summer in order to sail further north to the Lofoten Islands, which, so he understood, 'contain the most splendid rocky peaks in Norway'. On the first part of this cruise from Kristiansund to the Lofoten Islands and back in 1927, the crew consisted of Stephanie and G.I.'s regular sailing companion George McKerrow; and on the second part, from Kristiansund to the west coast of Scotland, G.I.'s brother Julian and his wife Edith replaced McKerrow. The whole cruise was described by G.I. in an article entitled 'Across the Arctic circle in *Frolic*, 1927' published in the *Royal Cruising Club Journal* (1927f). The article consists of the log kept by G.I. during the cruise, which is too long for reproduction here, and the following shorter account of interesting general aspects of the cruise:

> The northern part of the west coast of Norway is known to yachtsmen through the medium of Mr. Lynam's delightful book, "Norway and the North Cape in *Blue Dragon II*". It was largely this book which inspired us to go there, and we found the reading of the "book of the words", as we came to regard it, a continual source of pleasure on our journey. Our voyage was carried out under different conditions from his, though we necessarily covered much of the same ground. We had to get up north and back to the south of Scotland in less than seven weeks. We were therefore unable to see more than a small fraction of what Lynam saw. The information which we can give can therefore be regarded only as supplementary to his.
>
> The coast from Kristiansund northwards to Rörvik struck us as being less interesting than any other part of the coast. The part from Rörvik northwards is beautiful. The higher hills were covered with snow, and on our passage up there were patches of snow right down to sea level.
>
> So long as one keeps to the recognised channels and has clear weather there is little difficulty in navigation. The whole coast is strewn with rocks, but there are so many marks that, provided one keeps one's place on the chart and never misses identifying the marks, one can hardly go wrong. On the other hand, this profusion of aids to navigation is itself an embarrassment. It is

not always easy to distinguish them apart, even with a good glass, which is essential for this trip. Except during thick weather we identified every rock which should have been visible from our course, and we only saw two or three which looked like rocks and were not marked on the charts or mentioned in corrections. This speaks well for the Norwegian authorities responsible, but it makes long passages in a short-handed vessel very tiring. In a power-driven vessel such vigilance would be less necessary, because most of the dangers which one would meet while keeping exactly on the course are marked. A sailing vessel, however, necessarily has to tack across the direct course, and must keep clear of dangers by continually taking bearings. The necessity for this constant vigilance, combined with the complete absence of any darkness, even at midnight, made it extremely difficult for me to sleep., and I found some of our passages far more tiring than, for instance, a passage that McK. and I made some years ago in *Sorata* from Plymouth to Oban, or than one from Burnham-on-Crouch round C. Wrath to Loch Laxford, which we made in 1924.

For a cruise of this kind *Frolic*, with her large sail area of 1,250 sq. ft., is very suitable, because she moves well in light airs. Her great quickness in stays makes her very useful for beating in narrow channels. The chief trouble in handling her came in jybing. Except in light winds the full strength of our crew was quite incapable of moving the main sheet at all when running. As a rule, of course, it was possible to luff for an instant in order to get the sheet in, but on two occasions, while sailing in narrow channels, I have been forced by a change in the wind into a position where I was close to the shore, and sailing by the lee in a strong breeze and unable to get clear for fear of a jybe. In both cases we avoided disaster by rolling up the mainsail till we could pull in the sheet, jybing, and then unrolling again. For this reason I do not think we should have been able to run safely in narrow waters without a roller reef gear.

The Norwegian charts are extremely good and well arranged. In the principal series on a scale of 1:50,000 the charts are numbered consecutively round the coast from 1 to 116. We took with us 4 to 83, and we used in this voyage Nos. 24 to 72. We found it convenient to store them rolled in bundles of 10, no indexing system being necessary, though it is convenient to have the key maps contained in the list of charts. This list may be obtained from any Norwegian bookseller, e.g. Jacob Dybwad, of Oslo.

There is little difficulty in getting fresh provisions, the towns being provided with admirable shops. Though tourists of all nationalities go in large numbers up the coast, both in huge liners and on the mail boats, we never met any except in the town of Aalesund. We saw no other yachts north of Kristiansund, and, apart from a few open boats, we saw only one vessel moving under sail alone during our northern trip. All the local fishing and general-utility craft are now motor boats. As far as we could gather with our imperfect Norwegian, we were the first yacht that the inhabitants of Moskenesö had ever seen. Curiously enough, the "Rednings" boats, i.e. the lifeboats which go every winter with the fishing fleet, have no motors; no doubt there is always plenty of wind when their services are required. These are fine ketches of about 30 tons designed by Colin Archer. During the summer months many of them are laid up in Kristiansund, and I am told by Mr. Leonnechen that they may be hired at a very cheap rate.

In travelling up the coast we were interested in noticing the sudden change which occurs about latitude 64° in the character of the flora. All down the Norwegian coast south of this latitude we found hardly any flowers which are not common in Scotland. At Kvalöen and further north we found a large number of alpine or arctic plants, such as dryas, viola biflora, primula farinosa, purple saxifrage, loiseluria, coral root orchis, moss campion, and many others familiar to gardeners in England.

It was the custom of the Royal Cruising Club to award several Challenge Cups each year for the best cruises, and the top award for 1927, the Club Cup, went to G.I. for his account of his cruise in 'Across the Arctic circle in *Frolic*, 1927' (1927f). The Commodore who presented the cups said: 'Mr Taylor's cruise ran to 1600 miles in a 19-ton cutter with a sail area of 1250 square feet – rather a lump of boat for a crew of two amateurs, even with the assistance of the wife of the owner – in the turbulent waters of the North. Upwards of 950 miles were sailed along the rock-strewn coast of Norway as far as 30 miles or so north of the Arctic circle, and 217 in the ocean passage from Norway to Shetland, and 158 more from there to Loch Eribol. Mr. Taylor writes with great modesty, and until one considers what he performed it would be easy to underrate his cruise which, in my opinion, is a fine one'.

Well, if after being skipper on all the enterprising and adventurous cruises described above G.I. can fairly be referred to as an 'amateur', one can only suppose the Commodore is using the word in a technical sense.

The award of this Club Cup pleased G.I. greatly, and he wrote, many years later: 'I must confess I am prouder of this award than any other in my career'. This, from a man who had been knighted, elected FRS, awarded the Royal Society's Copley Medal and the US Medal for Merit, and appointed to the Order of Merit!

The only later cruise in *Frolic* of which there is a detailed record was in the summer of 1931, and was a cruise with G.I.'s brother Julian, his old sailing partner George McKerrow, and an aircraft designer 'G.H.' as crew. G.I. wrote it up for publication in the *Royal Cruising Club Journal* with the title 'Round Ireland in *Frolic*' (1931e). There is also a later reference to a second cruise off the coast of Ireland in *Frolic* in the summer of 1935, but no log is available. So far as can be inferred from letters to Stephanie written diligently by G.I. while he was away from home, G.I. sailed in a yacht called *Herga* to Estonia and to Finland in August 1936; he had sold *Frolic*, perhaps because it was too large for his needs. It seems that Stephanie's last cruise with G.I. was in the late twenties.

G.I. resumed sailing after World War II, and there is a hand-written log, which is not easily deciphered, for a cruise in a newly acquired yacht, *Eiver*, described as 'an auxiliary gaff-rigged sloop of 24 feet waterline length and $8\frac{1}{2}$ feet beam', in June 1948. The preface to the log reads as follows: 'During our cruise of 22 days *Eiver's* crew consisted of Dr A.A. Townsend[1] who by his imperturbability makes up for whatever he may lack in nautical experience, and myself. Our plan was to sail from Burnham to Brixham, there to join the marine meeting of the Cambridge University Cruising Club.[2] In the two remaining weeks of our available time we hoped to visit Brittany'.

It appears to have been an enjoyable and interesting cruise, notable

1. Distinguished for his contributions to turbulence research as well as for his imperturbability. Alan Townsend and I were colleagues who came to Cambridge from Australia to work for PhDs on turbulence under G.I. in 1945; see chapter 14.
2. Of which G.I. was President from 1946 to 1960.

for the friendliness of the French people in Brittany whom they met during their forays ashore.

Thereafter two more yachts are mentioned in letters home, *Cloche d'or*, in which G.I. sailed round the Isle of Wight in the summer of 1951, and *Guiding Light*, in which G.I. sailed in some part of the waters surrounding the British Isles in the summer of each of the years 1952, 1953, 1954, 1956 and 1957.

Mention has been made in chapter 3 of the influence that the Christmas lectures by Oliver Lodge at the Royal Institution in 1897 exerted on G.I. as a boy. When in due course G.I. himself was invited to give a 'course of six lectures adapted to a juvenile auditory' at the Royal Institution at Christmas 1936, he chose to lecture on 'Ships'. He prepared separate lectures on floating, ancient ships, yachts and native craft, models and ship designing, navigation, and mechanically propelled ships, and assembled a large collection of slides which suggest the lectures were fascinating for children of all ages. Here is G.I.'s explanation of Archimedes' Principle in simple terms which was devised for these lectures and which turns up in a delightful little speech expressing his thanks to the Institution of Mechanical Engineers for awarding him the 1965 James Watt International Gold Medal: 'I suppose that people receiving medals don't as a rule quite know what to do with them but I have been given a gold medal for which I have actually found a practical use. Two years after I received the Royal Society's Royal Medal I had to give lectures on ships to children at the Royal Institution and this involved explaining the Principle of Archimedes.[3] I therefore told the children the famous story of how the King of Syracuse gave some gold to a goldsmith and asked him to make a crown with it. When the crown was made the king was not satisfied because he suspected the goldsmith had debased the metal. He asked his kinsman Archimedes if he could prove the purity of the metal and it was while thinking about this problem in his bath that Archimedes suddenly discovered his Principle

---

3. G.I. got a great deal of pleasure out of his understanding of Archimedes' Principle. Another, and less familiar application of this Principle, is described in Appendix A.

and leapt out of the bath shouting 'Eureka'. To illustrate the scientific side of the story I had a brass crown made of exactly the same weight as my Royal Medal and weighed crown and medal in opposite pans in a balance in air. Of course they weighed the same. I then weighed them in water and the gold medal was much heavier than the brass crown'.

G.I. said in later years that he 'found the children the most stimulating audience he had ever addressed.' An audience of curious children like G.I. at the same age probably brought out the best in G.I. as a teacher. As a university teacher required to be more systematic and careful in his exposition he was less successful.

The C.Q.R. anchor

In conclusion of this account of G.I.'s sailing activities we may note his remarkably imaginative invention of an entirely new type of anchor, especially suitable for yachts and seaplanes and other craft for which weight is an important consideration. As usual the story is best told by G.I. himself. In 1934 he published an article on 'The holding power of anchors' in the *Yachting Monthly* (1934h) in which he discusses the action of anchors generally, and then in 1971 he wrote a shorter retrospective article entitled 'The history of an invention' for the student magazine *Eureka* (1971c) at Cambridge. The shorter article is reproduced in full as follows:

> In 1923 I bought the 45 foot yacht *Frolic* which weighed 20 tons and drew 8 ft. 3 inches of water. Her big anchor weighed 120 lb. While winding up the anchor the sails were not capable of controlling the boat until the anchor was nearly up to the surface, and when anchored in 10 fathoms or more close inshore with an onshore wind the effort involved in winding up the anchor to get under control before drifting ashore was too much for me. This and some problems connected with the anchoring of seaplanes provided the incentive to think about the design of lighter anchors.
>
> The earliest anchors were simple stones so that the ratio holding power/weight, H/W, was less than the coefficient of friction measured in air, usually less than 1. The Greeks realised that a much bigger H/W could be attained by using a hook which would dig into the ground and they, or their

contemporaries, invented the stock, that is the long bar at right angles to the plane of the hook which prevents it turning out of the ground once it is in. Since the stock would hold the fluke (that is the bent up part of the hook) pointing upwards if it fell that way and so prevent it from acting, it was necessary to add a second hook on the opposite side of the shank, thus making the anchor symmetrical about two planes through the shank. This second hook is necessarily at such an angle to the shank that it prevents the first from dragging the shank downwards. For this reason the high values of H/W which could perhaps be attainable by a single hook cannot be had with a traditional anchor.

My problem was therefore to think of a way in which a single hook without a stock could be made to dig into the seabed whichever way it fell, and be stable when pulled horizontally below the surface. The solution I came to is shown in the sketch [figure 8.4]. The shank A is hinged to the fluke by a pin C whose axis is shown as the broken line CE. The blades D and J are nearly portions of circular cylinders with a common generator FG. The sketch shows the anchor seen from above and lying as it falls with A, J and G on the ground. When the chain pulls, the point G begins to dig in because it is aiming obliquely downwards, and the lateral pressure turns the blades further downwards because the centre of lateral pressure is ahead of the line CE. As the blade buries itself the centre of lateral pressure moves backwards and when it passes the line CE the direction of rotation of the blades about the pin C reverses; and after

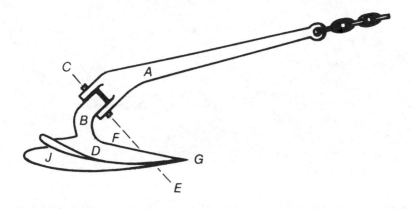

Figure 8.4 *Sketch of the anchor showing the lettering referred to in the text.*

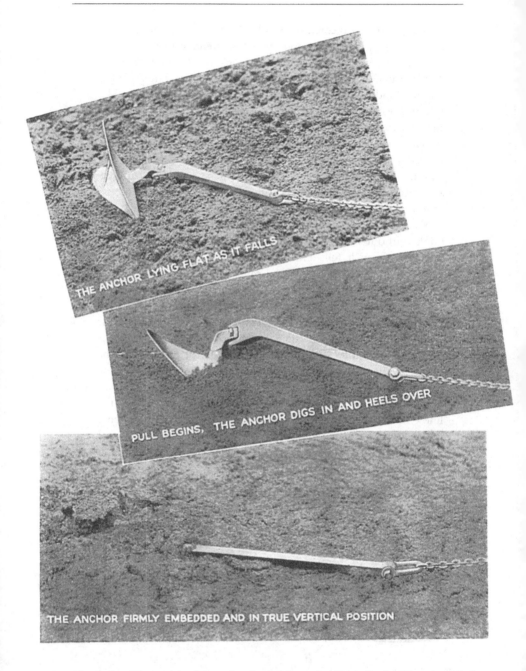

Figure 8.5 *Photographs of the anchor showing the way in which it digs itself into the sea-bed like a plough-share regardless of the way it falls.*

dragging a short distance the anchor assumes a position where the plane of symmetry is vertical. In this position the blades can pull the shank into the ground. Also the anchor is stable when pulled with horizontal shank and blades under the ground, for if it rolled slightly so that the blade J was lower than the blade D, J would be in ground which was deeper and therefore more difficult to move than that round D. Then the blades would rotate about the pin in such a way that the point G turned downwards and the anchor would return to the symmetrical position. By experimenting with a model on a sandy beach I found that the anchor could be towed in a circle keeping under the surface. When I dug up the blades while performing the experiment I found the blades banked over just like an aeroplane when it makes a turn, but remained symmetrical when pulled in a straight line (see figure 8.5).

The maximum value of H/W varied with the nature of the seabed, and in some grounds H/W was greater than 100 which is four or five times as great as that attainable with the traditional stocked anchor and 20 times that of the stockless anchors which all big steamships carry.

After inventing the anchor, I, together with my friends George McKerrow and W.S. Farren, set up a small company to make them for our sailing friends. Farren undertook the making of drawings suitable for supplying to a manufacturer, McKerrow arranged the marketing and I gave the invention. We called the company 'The Security Patent Anchor Co.' and would have liked to put the word 'secure' on the anchor, but it is not allowable to register a common word in that way, so we compromised and called our product 'C.Q.R.' For a long time afterwards people asked me what the Q stood for.

The history of this small company is instructive. It was founded in 1933–4 and had only begun operations and declared a small dividend of 5 per cent on the minute capital of £900 in 1939 when war broke out. Since the anchor was only intended for yachts we expected our operations to close down but soon the Admiralty started ordering them for their torpedo boats and George McKerrow who had taken on the job of managing director was kept busy through the war. Finally they were copied by Lord Mountbatten's combined operations group and used to anchor the floating 'Mulberry' harbour from which the Normandy landings were launched in 1945.

When the war broke out the Government passed antiprofiteering legislation which made it illegal for companies to increase their dividends and limited the fees payable to directors for attending to the business of companies. This was hard on a company like ours which had only just reached a stage at which there were any profits at all and it had effects which cannot have been intended. For instance, the anchor began to be used for purposes for which it had not originally been designed and the company was asked for advice. In some cases I made experiments to supply an answer. In other words I acted as a consultant, but as I was a director I was not allowed to charge for my advice. Nor was I allowed to resign from being a director to become a consultant, because according to a legal adviser that would have been regarded as a shady transaction, even though I received in fact no payment as a director. Since the capital of the company was only £900 such fees would have been minute anyway.

After the war a grant was made to the company for the use of the patent during the war though we did not ask for it. The grant was small since I had given the invention to the company, so that it had no assignable value. Even so most of it would have been absorbed as income tax if it had been distributed. In any case the patent had only a few more years of life and all of us had other things to do, so we sold the company with the patent included in its finances to a firm which could use the grant for development. It still makes C.Q.R. anchors.

It was characteristic of G.I. that the traditional practice and accumulated wisdom of centuries concerning anchors should not have deterred him from looking at the problem afresh. I am sure he was not motivated by a wish to show he could do better; it was simply that independent thought was instinctive for him. The C.Q.R. anchor is a triumph of geometrical and mechanical imagination. Its holding power for given weight is quite simply the best available, and it is used by small-boat sailors all over the world. Another view of the significance of the new design for its creator was noticed by Edgar Adrian, a close life-long friend of G.I., who remarked to him at a conference party in the Cavendish Laboratory in 1934 that 'I think you get more fun out of your work than anyone else I know.'

CHAPTER 9

The golden years as Yarrow Research Professor between the wars

I n 1923 G.I. was appointed to a Royal Society Research Professorship, one of two made possible by a benefaction from Sir Alfred Yarrow. This relieved him of the need to undertake teaching and other duties; indeed the terms of the professorship did not allow him to give more than a certain small number of lectures each year ('paid provided he does no work' as Rutherford put it) and he was able to devote himself totally to his research. Some scientists find full-time research a little too much of a good thing, and welcome some other duties as a diversion when progress is slow or painful, but not so G.I. Scientific enquiry was as natural for him as breathing, and his character and his inclinations were perfectly adapted to the pattern of life now open to him. G.I. was also no doubt glad to be free from teaching responsibilities. He had been a lecturer in mathematics at Trinity College since 1919, and four years were enough to show that his heart was not in teaching. I have heard it said that when the news of his appointment as the Yarrow Research Professor reached him at Cambridge in the middle of giving a lecture to undergraduates, he put down the chalk there and then.

G.I. naturally elected to hold his research professorship at Cambridge, and he continued to do his experimental work in the Cavendish Laboratory with the consent of Sir Ernest Rutherford and later his successors as Cavendish Professor, Sir Lawrence Bragg and Sir Nevill Mott. The Yarrow Research Professorship carried with it some money for supporting services, and G.I. was able to employ a technician, Walter Thompson, who remained with him for about 40 years. Thompson made all G.I.'s apparatus, usually from no more than rough pen sketches, and his devoted help and skill in trying different experimental schemes, searching for the right materials, and generally making things work, undoubtedly contributed to the success of G.I.'s research. I think Thompson probably sacrificed opportunities for promotion to higher-grade jobs in order to remain with G.I.; if he did so, it would have been pride in his position as assistant to a great scientist that kept him there.

G.I. held his appointment as Yarrow Research Professor from 1923 until he reached the retirement age in 1951, but it is convenient to consider first the period 1923 to 1939 in view of the change of activities imposed by World War II. In these years between the two wars G.I. was at the height of his powers and he had the opportunity and the facilities to use them fully. His investigations in the mechanics of fluids and solids covered an extraordinarily wide range, and most of them exhibited the originality and insight for which he was now becoming famous. It would be quite impossible to describe in this book all the contributions made by G.I. during this period, and only those of special significance will be singled out for mention. The list of G.I.'s publications at the end of the book provides a record of research over the whole of his life which is believed to be very nearly complete. The nature of G.I.'s thinking and the style of his work had been established by 1923, and have perhaps been conveyed already here, so that there may be less need for description of all his later investigations.

The two major themes: plasticity and turbulence

In this period from 1923 to 1939 G.I. devoted a significant part of his research effort to the mechanics of solid materials. About 20 per cent of all his scientific papers lie in this broad field, and virtually all the others lie in the companion field of mechanics of fluids. The range of scientific problems in solid mechanics that he investigated is less wide than that in fluid mechanics, but the originality and depth of his contributions to mechanics of solids are no less remarkable. He had a particular interest in the mechanism of plastic deformation of metal crystals which had been aroused at Farnborough in 1914 by his development with A.A. Griffith of the soap-film analogy[1] as a

1. The equations which represent the torsion of an elastic bar are of exactly the same form as those which represent the displacement of a soap film due to a slight pressure difference across its surface, the edge of the film being a hole in a flat plate of the same shape as the cross-section of the

method of measuring the torsion of an elastic bar of any uniform cross-section (1917a, 1917b, 1918a). In the course of this work they noticed that large shear stresses occur at internal corners. This seemed to shed light on the causes of the many failures in keyed crankshafts which were occurring at that time, and they used the soap film analogy to calculate how much internal corners must be rounded off in order to reduce stress concentration to tolerable values. It turned out, however, that in practice shafts with only slightly rounded internal corners, for which the calculated stress concentration was still large, often did not fail. It seemed therefore that the material must be behaving plastically.

The general question of the nature of plastic deformation of metal crystals and the relation to their strength occupied G.I.'s mind throughout the decade 1924–1934 in particular and he made many important advances. But plasticity was only one of two major themes in G.I.'s research in this astonishingly fertile period between the wars. The other was turbulent flow of fluid, an old preoccupation dating from his measurements of wind velocity near the ground while holding the Schuster Readership in Dynamical Meteorology in 1912. To each of them G.I. returned again and again, trying out new ideas, making new observations and testing new hypotheses. There were other topics on which he wrote several papers during the period but in no case is the number of papers near that relating to one of these two major themes. I have therefore assigned a whole chapter to each, that for plasticity being chapter 11 and that for turbulence being chapter 12. It turned out that the needs of World War II effectively put an end to G.I.'s research on plasticity and on turbulence, and he did not take up again seriously either of these two themes after the war. The descriptions of G.I.'s work on these topics given in chapters 11 and 12 consequently include his last papers in these fields.

(footnote 1 continued from p. 118)
bar. The method was technically useful in pre-computer days because there is no restriction on the shape of cross-sections for which the analogy holds. Although G.I. did not know it at the time, Prandtl had already suggested use of the soap-film analogy.

In the remainder of this chapter I shall describe briefly some of G.I.'s contributions to topics other than plasticity and turbulence during the period 1923–1939. These topics all lie in fluid mechanics.

Stability of shearing flow of stratified fluid

As a meteorologist G.I. was well aware of the effect of cooling of the ground at night on the wind. The wind near the ground drops, evidently as a consequence of suppression of the turbulent transfer of momentum by the vertical density gradient. To the mathematician this suggests the problem of the behaviour of small disturbances to steady flow of inviscid fluid in which the density and (horizontal) velocity vary with height above a horizontal plane representing the ground. This problem caught G.I.'s interest first in 1914, and the progress he was able to make formed part of the essay for which he was awarded the Adams Prize at Cambridge in 1915. The title of the essay is 'Turbulent motion in fluids' and it will be referred to again in chapter 12 on turbulence. (At that time it was common practice to regard considerations of stability of steady flow systems as part of the study of turbulence, presumably on the grounds that an unstable steady flow system usually becomes turbulent. However, nowadays we are less hopeful of being able to relate the growth of small disturbances in an unstable flow to the resulting turbulent flow, and the two areas are studied separately.) G.I. delayed publication of his work on the stabilizing effect of density stratification, at first on account of the war and later because he hoped to be able to undertake experiments designed to test the results. However, in 1930 or 1931 his colleague Sydney Goldstein told him he was working on similar problems and persuaded G.I. to publish his theoretical work alone (1931b).

The difficulty of this stability problem lies in the fact that the undisturbed vertical distributions of velocity and density are not given, and must be chosen with both physical significance and mathematical tractability in mind. The most appropriate choice for the velocity distribution appeared to be a uniform shearing motion in fluid of infinite vertical extent, whereas for the density there were several possibilities. The results obtained by G.I. may be summarized by the statement that a steady uniform shearing flow of stratified fluid is stable to small disturbances sinusoidal in the (horizontal) x-direction when

$$\frac{g|\Delta\rho|}{\rho L(dU/dz)^2} > c,$$

where $\Delta\rho$ is the decrease in density over a layer or central region with length scale L in the vertical z-direction and c is a number of order unity which has different values for different undisturbed vertical distributions of density. The dimensionless parameter on the left-hand side of this inequality is a measure of the ratio of the stabilizing buoyancy force on the fluid to the destabilizing inertia force, and later became known as the Richardson number.

G.I.'s relatively simple calculation for particular distributions of ρ and U as functions of z was a strong stimulus to mathematicians to consider the stability problem in more general form, that is, for arbitrary functions $\rho(z)$ and $U(z)$. Analysis of the behaviour of small disturbances in this system in which both gravity waves and vorticity waves co-exist is difficult but after much work a quite strong result was established many years later,[2] namely, that a flow is stable if the *local* Richardson number (as above but with $d\rho/dz$ replacing $\Delta\rho/L$) exceeds $\frac{1}{4}$ throughout the flow. Characteristically G.I. had conjectured this result from his analysis of the case in which dU/dz and $d\rho/dz$ are constant in a semi-infinite fluid with one plane boundary.

In a companion paper (1931c) G.I. considered the related but mathematically quite different problem of determining the critical value of the magnitude of the density gradient above which statistically steady turbulent flow with uniform mean shear cannot persist. Maintenance of the turbulence is possible only if the rate of working of the Reynolds stress exceeds the rate of gain of gravitational potential energy, and with simplifying assumptions about the turbulent transfer this is equivalent to

$$\mu_m\left(\frac{dU}{dz}\right)^2 > \frac{g\mu_s}{\rho}\left|\frac{d\rho}{dz}\right|,$$

2. J.W. Miles, *J. Fluid Mech.* **10**, 1961, 496.

where ρ is now the mean density and μ_m, μ_s are the (hypothesized) eddy transfer coefficients for momentum and a conserved scalar quantity respectively. Results of this type were also obtained by Richardson[3] (who acknowledged the earlier work by G.I. in his Adams Prize essay) and by Prandtl.[4] The three criteria for persistence of the turbulence differ mainly as a consequence of different ways of representing the eddy transfer, and none of them can be correct to better than order of magnitude.

Only G.I. compared the theoretical results with observations. He found some data obtained by Jacobsen[5] at two places in Scandinavia where a current of fresh water flows over a body of salt water. By measuring the vertical distributions of mean density and mean velocity Jacobsen had been able to infer the values of the transfer coefficients. The results were surprising. The values of μ_s/μ_m were all much less than the value of order unity expected from previous observations of transfer in a turbulent flow of fluid with uniform density, and the above criterion for persistence of the turbulence was satisfied. It appears that a kind of turbulence in which momentum, but not salt, is transferred vertically at the expected high rate can exist in a stratified fluid. G.I.'s paper (1931c) is partially about internal gravity waves, and it is probable that he was thinking of this 'kind of turbulence' as made up wholly or in part of superimposed randomly oriented internal waves which contribute little to the transfer of salt. He did not take the matter further, and it remains in need of more work.

Drops and bubbles and particles

Many of G.I.'s experiments involved the use of humble ingredients, and there were several, during this inter-war period, which required little more than the presence of two different phases, one of them being in the form of a discrete liquid drop or gas bubble or solid particle and the other a surrounding fluid. For G.I. the interest usually lay in the interplay of the various contributions to the stress at the

3. *Proc. Roy. Soc. A* **97**, 1920, p. 354.
4. *Vorträge aus dem Gebiete der Aerodynamik und verwandten Gebeiten*, Aachen, 1929.
5. *Beitrag zur Hydrographie der Dänischen Gewasser*, 1913.

interface in different circumstances and the associated phenomena. There is not much coherence in this little group of papers, although all the related theory is based on the assumption that the Reynolds numbers of the flow inside and outside the drop or bubble or particle are small so that fluid inertia forces may be neglected. What we see from the papers is G.I.'s unerring ability to spot interesting scientific possibilities in the interaction of such simple components.

Consider for instance the note in which he and his Cavendish colleague C.T.R. Wilson described observations of the change of shape of a 'soap-bubble' as the strength of a uniform electric field is gradually increased (1925e). The case of a charged or uncharged water drop in a uniform electric field is of more direct meteorological interest, but they perceived that it is easier to control the conditions in the case of an uncharged soap-bubble. They placed a half soap-bubble of known volume on the upper surface of the lower of two horizontal wet aluminium plates, and maintained a known potential difference between the two plates. The bubble was seen to elongate vertically as the field increased and to pass through a strange sequence of steady shapes near the highest point in particular. This tip became more pointed and conical, and above a certain critical value of the field the tip ceased to be stationary and instead vibrated rapidly in the vertical direction, throwing filaments and drops of liquid off the tip. Some of the shapes are shown in figure 9.1. Figure 9.1(d) is especially interesting because it shows the almost perfect cone into which the top of the bubble is drawn at the moment when the ejected filament is just formed. The conical tip shape intrigued G.I., and almost 40 years later he returned to the problem and was able to obtain the cone angle theoretically. The phenomenon of 'tip streaming' also made another appearance later.

A useful piece of work by G.I. which, untypically, was wholly theoretical concerned the effective viscosity of a body of fluid containing numerous small particles (1932f). Einstein had shown some years before[6] that the effective viscosity exceeds that of the pure fluid (μ)

---

6. A. Einstein, *Ann. Physik* **19**, 1906, p. 289 and a correction to that paper in *Ann. Physik* **34**, 1911, p. 591.

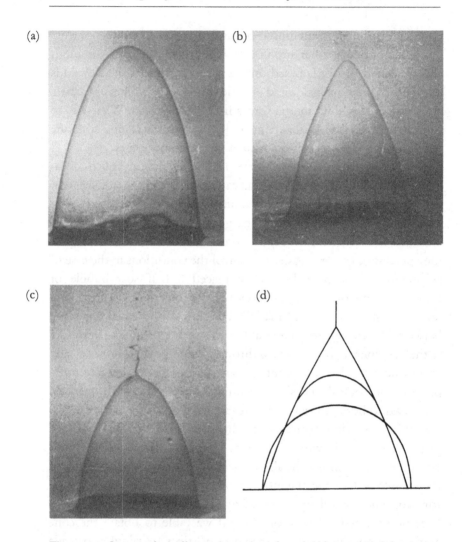

Figure 9.1 *Stages in the oscillation of an uncharged soap bubble in a vertical electric field. In (c) a filament is being thrown off the tip of the soap-bubble, and (d) shows tracings of three observed shapes, one of which is an almost perfect cone which forms at the instant at which the filament is ejected. (From 1925e)*

by the amount $\frac{5}{2}\phi\mu$ in the case of rigid spherical particles, where ϕ is the fraction of the total volume occupied by the particles. ϕ was here assumed to be small, so that each particle was effectively alone in infinite fluid. G.I. spotted that, for sufficiently large surface tension, liquid drops or gas bubbles would remain spherical, in which event the mathematical analysis needed for deduction of the effective viscosity of the dispersion would parallel that used for rigid spherical particles. In particular he noted from Einstein's analysis that, for any kind of spherical particle, the contribution of one particle to the effective viscosity is proportional to the coefficient of r^{-3} in the expansion of the disturbance to the fluid velocity in inverse powers of distance r for a particle in an ambient pure straining motion (as would be expected from the fact that this coefficient reflects the strength of the force dipole representing the particle). For a dilute dispersion of rigid particles the non-dimensional part of this coefficient is $\frac{5}{2}\phi$, and the excess viscosity of the dispersion is $\frac{5}{2}\phi\mu$; and for spherical liquid drops or gas bubbles of internal viscosity μ' the non-dimensional part of this coefficient is

$$\frac{5}{2}\phi\frac{\mu'+\frac{2}{5}\mu}{\mu'+\mu},$$

whence this (when multiplied by μ) is the excess viscosity due to the presence of the drops or bubbles. Thus in this easy way, which also avoided getting mixed up with the non-absolutely convergent integrals formally present in Einstein's proof – although only formally, because actually only differences between two such integrals occurred and these *were* absolutely convergent – G.I. was able to derive the effect of the internal circulation in the liquid drops or gas bubbles on the effective viscosity of a dilute dispersion.

The latest and undoubtedly the most influential of this small group of papers on drops and bubbles was concerned with the deformation and rupture of a liquid drop owing to motion of a second liquid in which the drop is suspended (1934 g), that is, with the way in which an emulsion of very small droplets is formed. G.I. constructed two pieces of apparatus for the generation of a representative flow field of the outer fluid, one consisting of four cylinders mounted at the corners of a square and rotating in opposite directions which generated a two-dimensional pure straining flow at the centre, and the other consisting of two endless loops of cinema film stretched between rollers which rotated and produced a simple shearing motion of the fluid between adjoining parts of the two loops moving in opposite

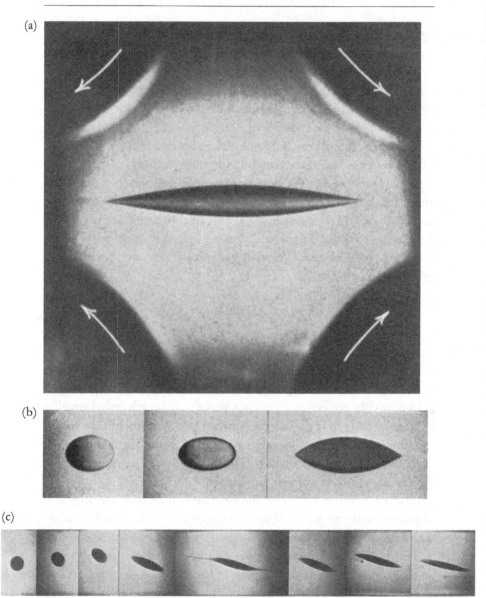

Figure 9.2 *Steady-state shapes of drops in a steady ambient flow field (1934g). (a), (b) Drops in the four-roller apparatus, with increasing flow intensities from left to right in (b); (c) drops in the parallel-band apparatus, with anti-clockwise vorticity and increasing flow intensity from left to right. The filament coming off each end of the fifth drop in (c) is a transient phenomenon.*

directions. For each of these two ambient flow fields a drop of a second liquid was placed at the centre of the flow field, and the shape adopted by the drop was studied as a function of the ratio of the viscosity of the drop $(\lambda\mu)$ to that of the outer liquid (μ) and of the capillary number $F = \mu\alpha a/\gamma$, where α is the ambient rate of strain of the fluid at the position of the drop, a is the radius of a sphere having the same volume as the drop, and γ is the surface tension. As already remarked the Reynolds numbers of the fluid motions inside and outside the drop were assumed to be small. Even so, the dependence on two dimensionless parameters for each of two flow systems gives rise to a formidable observing programme. No previous observations or theory were known to G.I, so he chose to study those parameter values which corresponded to interesting drop shapes, in particular those that were a prelude to break-up of the drop. Figure 9.2 shows some representative steady states.

Since this work by G.I. in 1934 there have been many other experimental, and, in recent years, numerical studies, and it is no longer necessary to digest the tentative qualified conclusions of the pioneer. The more complete understanding reached by 1984 has been summarized as follows:[7]

1. Drop deformation and break up is promoted primarily by the straining motion in the outer flow. It is inhibited by the vorticity in the outer flow.
2. For globular drop shapes, surface tension acts as a restoring force resisting burst, but if the drop becomes elongated, then surface tension may promote rupture.
3. Low-viscosity drops can attain highly extended stable shapes and require very strong flows to break them.
4. High-viscosity drops are pulled apart by modest extensional flows (though they take a long time to break), but cannot be broken by flows with significant vorticity, however strong.
5. The history of the outer flow can be important in determining whether or not a drop breaks: rapid changes in flow strength can provoke subcritical rupture. The size and number of

7. See J.M. Rallison, *Ann. Rev. Fluid Mech.* **16**, 1984, pp. 45–66.

satellite drops produced by rupture depend on both the flow type and its history.

G.I. had formed a simpler picture of bursting and break-up (item 5 above). He wrote, 'The act of bursting is always an elongation to a thread-like form. When this thread breaks up it degenerates into drops which are of the order of one hundredth of the size of the original drop. This seems to be related to the known fact that when an emulsion is formed mechanically it contains drops which cover a very large range of sizes.' The precise way in which the elongation to a thread-like form (tip-streaming again) took place was unclear, and has remained so, although 30 years later he was able to make some theoretical progress with the situation in which the tip of a drop is a narrow cone and the outer fluid flow near the tip of the drop is axisymmetric (1966a).

High speed flow of gas

The knowledge of aeronautics that G.I. gained at Farnborough during World War I made him valuable as an adviser on government committees, and the list of subcommittees and panels of the Aeronautical Research Committee of which he was a member shows that he maintained a close connection with aeronautical research throughout the period between the two wars. I presume that it was through one of these committees that his attention was drawn to problems of flow of air at such high speeds that compressibility effects are important, for his first paper in this area (1928f) was written in collaboration with a Mr C.F. Sharman who was enabled to work with him for a year by a grant from the Air Ministry. The paper sprang from the realization by G.I. that the equations describing steady two-dimensional irrotational flow of a compressible fluid are identical with those describing the flow of electric current in a plane conducting sheet of variable thickness. In the former problem the fluid density is related to the fluid velocity at the same point by Bernoulli's equation, whereas the analogous quantity in the latter problem, the sheet thickness, may be chosen arbitrarily. G.I. perceived that this analogy enabled a problem of steady flow of air at high speeds to be solved by an iterative set of measurements of electric current flow through a shallow layer of electrolyte in a tank with paraffin wax at the bottom, the bottom being carved to the depth distribution required by Bernoulli's equation after each measurement. The iteration is similar, although not quite identical, to the mathematical successive-approximation procedure devised by Rayleigh in 1916 for the problem of fluid flow past a circular cylinder. Taylor and Sharman showed that the method gives results of adequate accuracy in a reasonable time, regardless of whether the boundary is of simple geometrical form, and was probably superior to any mathematical method. However the limitation to two-dimensional flow is a handicap to aeronautical applications and the method seems not to have been much used.

G.I. appears to have been interested in the analogy partly as a workable method of solving practical problems and partly as a means of investigating the conditions under which continuous irrotational flow ceases to be possible. Rayleigh had speculated that irrotational flow past a circular cylinder ceases to exist when the ratio of the speed in the free-stream to the local speed of sound reaches unity, but the electric-current method in fact failed to converge for values of this ratio above 0.5. Taylor and Sharman noted that locally supersonic flow makes its appearance first (at the side of the cylinder) also when the free-stream Mach number is about 0.5, and this led naturally to the new hypothesis that continuous irrotational flow is possible only when the maximum speed in the flow field is less than the speed of sound at the same point. As the authors noted, this hypothesis is suggested by experience with a Laval nozzle, but its applicability to flow past bodies had not previously been appreciated. In two later papers published by the Aeronautical Research Committee (1930b, 1930c) and in a lecture to the London Mathematical Society (1930d), G.I. searched for some theoretical argument underlying this hypothesis but it was elusive and the best he could do was to show that when the maximum velocity in the flow field is supersonic, small corrugations of the boundary produce disturbances of finite amplitude in the stream. He had put his finger on the essential point, and further elucidation of the questions had to wait on application of the theory of second-order partial differential equations of hyperbolic type.

G.I. turned his attention next to cases of supersonic flow with shock waves and made two useful contributions. The first (1932e) was a straight-forward comparison of theoretical results for the forces on a thin biconvex aerofoil moving at supersonic speeds with some measurements made in a wind tunnel by Stanton. The theory was due to Ackeret, who showed in 1925 that the flow could be regarded as a combination of a uniform stream, weak shock waves attached to the leading and trailing edges, and rarefaction waves attached to each point of the curved surfaces, and G.I.'s objective was simply to show its applicability to aeronautics and to quantify the conditions under which the leading shock wave is, as assumed in the theory, attached to the leading edge. The second (1933b,c) was a more substantial investigation with J.W. Maccoll of the flow due to a conical body moving at supersonic speed in the direction of its axis with an attached shock wave. Busemann had pointed out in 1929 that the flow behind the attached conical shock is irrotational and 'conical' (i.e. the velocity, pressure and density are constant on coaxial cones with the same vertex), although without giving any details or results, and Kármán and Moore[8] had just published an approximate theory for thin axisymmetric bodies. Taylor and Maccoll calculated the irrotational flow solution completely, showed that the theoretical value of the angle of the attached conical shock agreed well with that observed on bullets at Woolwich Arsenal, found good agreement between the calculated surface-pressure distribution and that observed on a conical body in the supersonic wind tunnel at the National Physical Laboratory, and confirmed the accuracy of Kármán and Moore's approximate solution for cones of sufficiently small angle (see figure 9.3). It was not an investigation which called for G.I.'s originality, but it was a definitive scientific examination of the supersonic flow past cones which has been much used.

Long gravity waves in the atmosphere

G.I. maintained his early interest in dynamical meteorology, and in 1929, in preparation for participation at the 4th Pacific Science Congress in Java, he began a study of long gravity waves in the atmosphere. This was apparently prompted by the fact that the Congress

---

8. *Trans. Amer. Soc. Mech. Eng.*, 1932.

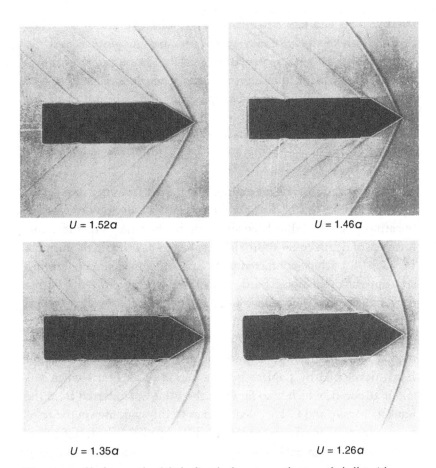

$U = 1.52a$

$U = 1.46a$

$U = 1.35a$

$U = 1.26a$

Figure 9.3 *Shadow-graphs of the leading shock wave near the apex of a bullet with a 60°-conical head moving at supersonic speeds. Theory is able to account for the angle of an attached conical shock wave and the air pressure at the surface of the bullet. The shock wave is detached when the bullet speed U is less than 1.46 times the speed of sound a. (From 1933 c, d)*

delegates were to visit the island of Krakatau, near Djakarta, where in 1893 there was a great volcanic explosion, the most violent ever recorded, which blew off half the island. The explosion generated a gravity wave in the atmosphere which barograph records showed travelled three times round the Earth, with a mean speed of 1046 ft/s. G.I. showed that the known general features of the distributions of wind velocity and temperature over the globe account for the shape that the Krakatau wave developed after passing once round the Earth (1929c). He also calculated the velocity of a long adiabatic gravity wave in a plane non-uniform atmosphere whose temperature decreases with half the adiabatic lapse rate up to a height of 13 km and is constant thereafter – a more realistic vertical distribution of temperature than that assumed in earlier work by Lamb – and found it to be within 2 per cent of the observed speed of the Krakatau wave (1929b).

This calculation was interesting, quite apart from the nice agreement with observation, because the speed of a long gravity wave in the atmosphere had thus been shown to be the same as that in a uniform sea of depth equal approximately to 35 000 ft (from the formula $U = (gh)^{1/2}$), which conflicted with some other evidence concerning atmospheric oscillations. Barographs show a persistent semi-diurnal oscillation, and the general view of geophysicists, following a suggestion by Kelvin, was that this must be due to resonant amplification of the disturbances from either the tides or the diurnal heating. The theory of free oscillations of a uniform sea of depth h on a spherical Earth developed by Laplace gives a period of 12 hours when h is about 26 000 ft, which is so far from the value of h inferred from the Krakatau wave and G.I.'s calculation as to throw doubt on the resonance theory. Geophysicists wondered (a) whether a short pulse generated by a sudden explosion would travel with the speed of a long wave and (b) whether radiation damping of disturbances to the atmosphere would be sufficiently strong for long period waves to have a velocity different from that of the short-period Krakatau wave and the adiabatic wave considered by Lamb and Taylor; but in a later paper (1932d) G.I. demolished both these escapes from the difficulty by showing that the breadth of the Krakatau 'pulse' is about 100h, quite large enough for long wave theory to be applicable, and that if radiation damping were strong enough to affect the wave speed it would also reduce greatly the resonant amplification.

The difficulty in accounting for the observed semi-diurnal oscillation of the atmosphere evidently nagged at G.I., for a few years later he made a calculation of the modes of (adiabatic) oscillation of an atmosphere with a non-uniform vertical distribution of temperature on a rotating globe (1936c). He was able to show that, contrary to what had generally been supposed, there are in general many modes of oscillation, and correspondingly many periods of oscillation, although for the particular cases of an isothermal atmosphere and an atmosphere in adiabatic equilibrium, which had been examined by Laplace and Lamb respectively, it happens that there is just one. This provided the resolution of the problem because a little later Pekeris found from G.I.'s analysis and up-to-date data for the vertical distribution of temperature in the atmosphere that there are in fact two modes of free oscillation with periods close to 12 hours; on the other hand, the estimate of the speed of a pulse like the Krakatau wave with dimensions small compared with the radius of the Earth was unchanged by the new analysis.

CHAPTER 10

Tour in the Far East

C haracteristically G.I. decided that his attendance at the 4th Pacific Science Congress in Java in 1929 should provide an opportunity to explore some of the wilder parts of Malaysia. After a three-weeks voyage by sea to Sumatra, G.I. and Stephanie visited many of the innumerable neighbouring islands, many of them volcanic, taking advantage of the guidance provided by the Congress organisers. G.I. also had ambitious plans to see tropical jungle, and he asked the authorities for permission to visit West Borneo and to try to cross overland to Sarawak. The Governor of West Borneo was sceptical about their chances of success and uncertain whether he should help. However, G.I. never lacked confidence, and in the event he and Stephanie travelled about 380 miles up the Kapoeas River on a small boat fixed alongside a steam tug; they then journeyed along smaller rivers in two dugout canoes with a party of dyaks until they reached lake country; and finally walked along narrow jungle paths. They entered Sarawak at a frontier fort where Rajah Brooke of Sarawak had arranged for more dyaks to transport them down the Batang Lupan river to Kuching, the capital of Sarawak, in a large dug-out canoe. (See the map in figure 10.1.)

G.I. made notes on the journey but they are difficult to decipher. Here is Stephanie's account, apparently prepared for delivery as a lecture with slides and a cine-camera film:

> The journey through Borneo which I am going to describe was undertaken by my husband and myself for the simple purpose of seeing a primitive people in a country which has, as yet, been but little affected by what we know as civilisation. Borneo is by no means an easy land to visit, at any rate for those who have no official business there, and I should like to say at the outset that this journey could never have been undertaken without the kind

134

Figure 10.1 *Sketch-map of Borneo.*

permission and help of the Governor-General of the Dutch East Indies and his officials, and of Rajah Brooke of Sarawak.

There are many reasons why it is not easy for the casual traveller to visit Borneo. In this huge island whose area is about four times that of the British Isles, there are no roads of any extent, no railways, no overland telegraphic or telephonic communications and no hotels, and a journey through the country necessitates a certain amount of preparation beforehand. The ordinary way of travelling is by river and our idea was to start from Pontianak, the Dutch port on the Kapoeas river, and to make our way right up into the interior, following the river as far as we could, then over the little known country of the border land between Dutch Borneo and Sarawak and so down the Batang Lupar river, through Sarawak and ultimately to Kuching, the capital, a journey of several hundred miles. In Java where we made our plans we consulted various kind Dutch friends, who told us that this was a possible route, though for several reasons we might not be able to make our way through. The most difficult part would be the border country between Dutch Borneo and Sarawak, the difficulty being that, if there has been too much rain huge areas are converted into swamps which it is impossible to traverse. As it was not possible for us to find out what was the condition of this part of our proposed journey, we had to trust that fortune would favour us.

In Batavia[1] we proceeded to lay in the necessary stores – tinned food, rice, soda water (this was very necessary as we naturally could not rely on the water of the country), camp beds, two large mosquito nets and as precautions against malaria large doses of quinine. We had taken with us from Woolworths a number of small articles suitable for camping utensils, and these made very useful presents later on for our Dyak porters who carried all this paraphernalia on their backs through the jungle. There were not quite enough of such presents to go round so I brought out a few silk scarves and other trifles of my own. I tried to explain that they were for their wives, however the men did not see the point of this at all – they tied them on to themselves, just anywhere and strutted about as proud as peacocks.

We started from Batavia on the 18th of June in a small and beautifully clean Dutch steamer. The Dutch captain spoke English well and was a very good friend to us, as will appear later. On June 20 we arrived at the entrance to the narrow dredged channel which leads up to one of the mouths of the Kapoeas river on which Pontianak is situated. It was really thrilling to catch our first glimpse of Borneo, and the wonderful jungle of palms and tropical trees came up to all our expectations.

Our arrival in Pontianak brought our first setback. By a most unfortunate mischance we had timed our visit just before the first official visit to Borneo of the Governor-General of the Dutch East Indies, who is a personage second in importance only to the Queen of Holland. Two very polite officials came on board and welcomed us in the name of the Governor, but they added that, as every scrap of their very limited accommodation was more than occupied by the officials who were assembling from all parts of Borneo to meet the Governor-General, they much regretted that there was absolutely nowhere we could stay while we made our arrangements to go up the river. This was indeed a setback. Pontianak is not by any means a health resort – it is right on the equator and lies on a swamp which might have been the scene of one of Conrad's novels. Sleeping out was not to be thought of, but we must sleep somewhere. Fortunately our kind friend the captain of the steamer, came to our rescue. He said that we might

1. Now known as Djakarta.

stay on board for that night, and in fact we stayed until the steamer went back to Batavia, after three days. All this time we were trying to find means of getting up the river and after the first day we began to feel that we should be glad to leave Pontianak. The boats all round us on the quay were loading rubber, which in its unprepared state has a most horrible smell, and that combined with the extreme heat, the mosquitoes and the many other smells of a tropical harbour made life somewhat uncomfortable. We eventually started, in a motor-boat belonging to a planter who lived up the river, but we had only progressed about a quarter of a mile upstream when the engine broke down and we were told that it could not be repaired till a new part could arrive from Java. The boat was towed ignominiously back to Pontianak and we began to think that our prospects of ever leaving the coast to say nothing of getting through to the other part of Borneo were very poor indeed. But our luck was about to change. The next day, after a really unpleasant night on board the motor launch, tied up to a Chinese boat that was none too clean and whose crew looked very much like pirates, we heard that a curious looking Noah's Ark-like object, which we now observed in the river, was in reality the government salt boat, and that it was about to proceed in the desired direction and was willing to take us as first-class passengers. With the greatest possible haste we transferred our baggage to this strange looking craft, and at last we left Pontianak for the unknown.

For four days and nights we progressed peacefully up the river, a distance of several hundred miles and every moment there was something new and wonderful to look at – inpenetrable jungle, orchids, monkeys and colourful birds. On the fourth day we arrived at the furthest outpost of Dutch rule, a little settlement called Semitau, where one Dutchman lived alone and administered the country round. It is a terribly lonely life in these outposts of government and one Dutch resident whom we saw on our way up to the country told us that all the time he was longing to get back to civilisation, but not so our Dutchman at Semitau. No man could have been better suited for life in the wilds. He loved the country and the people and knew how to treat them and he was thoroughly happy and contented though he was the only white man in 10,000 square kilometres of surrounding country.

We were astonished and grateful at the completeness of the arrangements he had made for us, and when we apologised to him for taking him away from the festivities connected with the Governor-General, he said that a journey into the wilds was far more to his taste than visits of the great and that he was really grateful to us for coming.

I will describe briefly the rest of our journey. For two days we travelled by motor-boat through an intricate series of lakes and channels – the most curious country imaginable. Trees grew out of the water in what was at that time a series of lakes, but at others a swamp, according to the time of year and the extent of the rains. This extraordinary country stretched for miles in every direction and it was this part of the journey which would have been impossible if it had been in a condition of swamp instead of being a vast series of lakes.

Our Dutch friend was a delightful companion. He told us what animals were to be found in that part of Borneo; orang-outangs, and several other kinds of monkey, crocodiles, a small rhinoceros, bears, also many kinds of snakes. Of the latter we had been warned of a small black one, so poisonous that death resulted from its bite within a few minutes. We naturally kept a careful look-out for him. We saw one such snake but few animals.

When the motor boat had penetrated as far as possible we and our goods transhipped into one of the two dug-out canoes, in which we were to continue our journey. By the time we ourselves, half the Dyak porters and half our baggage were in one of the canoes it felt fairly unstable and my mind turned to thoughts of crocodiles for we had seen the tails of several disappear into the water at our approach. But there was something far more exciting to occupy our minds than crocodiles, for now we were for the first time at close quarters with the Dyak porters and we could observe them without appearing to stare at them. They were attractive looking small people of a pleasant brown colour, most of them had very long hair, straight or wavy, and their slight costumes consisted of a loin-cloth, necklaces, white bone armlets and earrings. We also were interesting to them. [See figure 10.2.]

After some hours travelling in our canoes we reached a place where an enormous tree trunk blocked the stream and it was impossible to proceed further by water. We all disembarked, the canoes were tied up to a tree to await the return of the Dyaks and the Dutch resident, and we continued our travels on foot. That night we slept in a wooden hut, a remnant of what in early and more disturbed days had been a fort. The hut was in a dilapidated

On the Dutch-Sarawak frontier, Borneo
July 1st 1929.

Figure 10.2

state, and a number of gigantic bats and other weird creatures had to be removed before our beds could be erected.

After our evening meal we had a party, the most delightful party of our lives. The Dyaks had gradually grown used to us, and as we ate our supper they squatted round and watched us eat. We were very sad that we could not talk to them, but our Dutch friend interpreted for us, and through him we asked them to dance. We had luckily acquired a Dyak musical instrument, one of the men could play it and another was an expert dancer and volunteered to dance the ceremonial dance which in the 'good' old days when head hunting was allowed led up to a head hunting expedition.

A flaring torch was set up in a clearing, and we all sat round in a circle and watched the dancer. All around us was the jungle and the cries of wild beasts mingled at intervals with the music. Suddenly there was a stir in the forest and two strange dyaks walked into the circle of light. They were two chiefs from villages some distance away, who had come to pay us their respects. We greeted each other and did what we could with signs and smiles. They presented us with a bottle of arak (a fiery rice wine) and an egg. We reciprocated with presents, among which some trays purchased from Woolworths were especially popular.

The next day we set off again early. We crossed the frontier into Sarawak at a fort called Labuk Autu, and entered Rajah Brooke's territory and here on the narrow jungle track we met a Dyak hunter armed with a long spear in the afternoon. Here our Dutch friend left us. I shall not say any more about our journey through Sarawak except that we had here the new and exciting experience of shooting rapids in a canoe, a rather nerve-racking experience at first. We travelled entirely by river and chiefly in a dugout canoe. We slept in all kinds of queer places, including a real fort surrounded by an ancient cannon with an armed guard outside the door.

Ultimately we reached Kuching, the capital of Sarawak, in spite of many prophecies that we should not get there. Here we were most kindly and hospitably entertained by Rajah Brooke's financial minister and by Rajah Brooke himself. We felt very deeply grateful for the welcome we received from kind people all of whom were complete strangers to us, and the time we spent in Kuching was among our happiest memories.

I feel that I cannot leave the subject of Kuching without saying how impressed we were by the wise and benevolent rule of the British Rajah of Sarawak. It would be too long a story to relate how this large territory was handed over, two generations ago, to Sir James Brooke, the first rajah, by the Sultan of Brunei. By this deed of gift a young Englishman became the absolute ruler of a land of vast tropical forests and of wild tribes of head-hunting Dyaks who were always at war with each other. From the first he set himself to rule for the good of the natives of the country. He would allow no foreign exploitation, no European traders were allowed to come in and extract the riches of the land, and the two succeeding rajahs have most worthily upheld their trust. As far as possible the customs of the natives have been respected; no new modes of life, clothes or customs have been forced upon them or even encouraged, and the result is the interesting spectacle of a native race, protected from harm from outside and developing along its own lines, with the help of what is good in our civilisation, and with nothing of the bad. The country is administered from Kuching by the Rajah and a small number of British officials who are personally chosen by the Rajah, and there are of course a certain number of residents in the interior. The rest of the government officials are all natives, either Malays or Dyaks.

I said just now that the customs of the country were not interfered with. But naturally this could not be so in the case of head-hunting. This practice, which consisted in lying in wait for a valiant man of an enemy tribe and cutting off his head, unless he succeeded in cutting off yours first, could naturally not be allowed to go on. It was however of great religious significance, and as such it was very dear to the Dyak heart and so very difficult to eradicate. However, now after nearly 100 years the practice is really stamped out. Isolated cases do still occur, but they are very rare. The dried heads however are still the greatest treasures of a Dyak campong and will become increasingly precious as the years go by.

A few words about the inhabitants of the country. The Dyaks are a pagan people, consisting of about 6 different groups, which vary in appearance, in customs and in religious beliefs. They are great travellers and make excellent porters for any exploring expedition. For their dwelling place, one whole village lives together in a communal house known as a 'long house', so that they are all under one roof. The long houses are built on high piles for protection from enemies and wild animals and consist of a long gallery which is common to the whole village and a number of separate compartments ranged along one side, one for each family. As the young people grow up and marry a new bit is added to the house to accommodate them. The sole entrance to such a house is a slanting notched tree-trunk up which one climbs to the front door, a mode of entry which takes a little practice if it is to be done gracefully.

The Dyaks live by hunting, and by growing just enough rice for their needs. They also collect the edible fruits which abound in the jungle and which at some times of year form almost their only food. They weave and dye their own clothes, make their own weapons and implements, and many of them are very skilful carvers in wood and bone. Their wants are few and their lives simple but they are a very happy people. Children are never punished – a serious talk from the head-man of the village is enough to quell the most riotous spirits. They are all very ready to laugh and to enjoy a joke. They are, in fact, a most cheerful and charming race. They are perhaps not as beautiful as their neighbours, the Balinese, but they are very enterprising and if only the omens are favourable will go anywhere.

Figure 10.3 *The group that climbed Mt Hodaka in August 1929: Geoffrey and Stephanie and members of the Kyoto University Alpine Club.*

Stephanie was inclined to be fastidious about personal matters, and she must have been sorely tried by the conditions under which they lived, but her account of the journey shows her genuine interest and her active participation in a new and very unusual experience. As for G.I., he could cope with any situation and would enjoy doing so.

This taste of an unfamiliar environment and culture was repeated almost immediately, for on their way home from Java by sea they stopped in Japan (in July 1929) and spent several weeks there on exploration away from the cities. Stephanie kept an interesting and detailed diary of their adventures, and there is a film taken with her ciné-camera. It is clear that they enjoyed immensely their introduction to the sights and sounds of rural Japan, visiting shrines, walking in the mountains in company with some friendly members of the Kyoto University Alpine Club (figure 10.3), and staying in traditional Japanese inns. And the ciné film showing Stephanie resolutely wading through a fast mountain stream with dress tucked into her bloomers gives a new meaning to active participation.

The whole trip lasted nearly six months (April to October 1929), and included visits to many places en route to their main destinations, such as Colombo, Singapore and Hong Kong. G.I. knew how to work hard at his research, and his play was just as enterprising and intensive. Work and play were equally attractive to him, and this was no doubt the key to his full life.

As I write this there comes news of the development of tourist facilities in Sarawak. An article in a national newspaper says: 'From mid-summer 1994, tourists will be able to visit long houses without compromising their personal comfort when Hilton International opens an up-country hotel at Batang Ai. And with the recent opening of its Damai Beach hotel, Holiday Inn can offer two-centre breaks combining the rigours of the jungle with the undeniable attractions of the sun, sand and the South China Sea.' G.I. would not have been pleased.

CHAPTER 11

Plasticity of crystalline materials

As already noted in chapter 9, G.I.'s interest in the plastic behaviour of metals was aroused by his work with A.A. Griffith on the failure of aircraft propeller shafts at the Royal Aircraft Factory at Farnborough in 1914. His interest lay dormant for some years, until he was 'inspired' (so he wrote later) by a paper read at a meeting of the Royal Society in 1922 by H.C.H. Carpenter and C.F. Elam.[1] Initially he adopted a macroscopic or continuum viewpoint, as would be appropriate for considerations of plastic flow in single metal crystals, and proceeded to an extensive programme of novel experiments which laid the foundations of the subject. Later he conceived the idea that the movement of 'dislocations' (localized faults in the regular arrangement of atoms in a crystal) is the microscopic manifestation of plastic deformation, and developed the quantitative aspects of this idea in several papers beginning in 1934.

I am not familiar with these fields of G.I.'s research, and have therefore sought help from knowledgeable colleagues. Professor Rodney Hill has kindly contributed a comprehensive account of G.I.'s work on the continuum mechanics of plastic deformation, and Professor Sir Nevill Mott has kindly written about G.I.'s theory of dislocations.[2] In the third section of this chapter there is G.I.'s own account of 'the early stages of dislocation theory' (1965b) written 30 years later. The whole chapter is an interesting story which provides an impressive testimony to G.I.'s versatility and in particular to his talents as an experimenter.

1. *Proc. Roy. Soc. A.*, **100**, 1921, p. 329.
2. These two pieces were initially written for inclusion in my Royal Society Biographical Memoir of G.I. Taylor (*Biog. Mem.*, **22**, 1976, pp. 563–633), and are reproduced here with slight modifications only.

Mechanics of plastic continua (by Professor Rodney Hill)

During the mid-1920s Taylor mainly directed his energies to eluci-
dating the macro-mechanics of plastic flow in metal single crystals.
For his experimental programme it was essential to have very large
specimens (typically 20 cm long and 1 cm across), capable of being
machined with precision to a shape permitting homogeneous loading and the accurate observation of dimensional changes. For the
growth of such large crystals he depended on the skill and experience of Miss C.F. Elam (author of the classical monograph:
Distortion of Metal Crystals, Oxford, 1935). As he himself wrote:
'Without her help the work could not have been done.' He relied
too on aid from many quarters in such matters as the design and
operation of X-ray apparatus, the construction of a variety of testing
machines, and the loan of gadgets of all kinds; his papers acknowl-
edge scores of names, those of Miss Elam, W.S. Farren, and A.
Müller being especially prominent. Taylor announced the first fruits
of his programme in a Bakerian Lecture (1923a). Thereafter the
results flowed steadily, over a widening range and with increasing
accuracy (1925b, 1926a, 1926b, 1927b, 1927c, 1928a). In essence
the principles he extracted are simple: (i) at a macroscopic level of
observation, face-centred cubic crystals deform plastically by a specific type of crystallographic shear, which effectively regenerates the
lattice; (ii) the shearing is activated when the corresponding component of applied stress attains a critical magnitude; (iii) this magnitude
is independent of the normal stress on the shearing planes, whether
this component be tensile or compressive; (iv) two of the specific
crystallographic shears occur simultaneously when the respective
shearing stresses are both critical; (v) body-centred cubic crystals
deform plastically by 'pencil glide' in specific lattice directions but
over non-specific planes. Before Taylor's experiments, principle (i)
had been merely *inferred* from X-ray evidence and from slip-line
markings on crystal surfaces; (ii) and (iii) seemed in consequence
plausible but had not been shown conclusively; (iv) and (v) were
shrouded in a fog of conjecture.

Taylor gave the first *proof* of all these mechanisms, in situations
moreover where surface markings are absent. He did not find it easy,
however, to convince contemporary metallurgists of the intrinsic

superiority of his general approach. In later expositions he observed
with some asperity (rare indeed) that other methods depended 'on
knowing the form which the answer will take before starting to
solve the problem' or on 'verifying particular hypotheses by special
measurements' (1927a, 1938a). Nobody now would question the
justness of these remarks. Taylor's approach was, in truth, completely general: as many dimensional changes were recorded (usually
three lengths and three angles) as would uniquely determine an arbitrary uniform deformation, unknown *a priori*. In a tension test, for
instance, square rods were taken from the crystals at orientations
given by X-ray analysis, and the lateral faces of each rod were
scribed so that its deformation could be checked for homogeneity.
In preference to the strain ellipsoid, Taylor located the quadric
cone of unstretched material lines; from this the deformation could
be conclusively demonstrated to be a simple shear whenever the
cone degenerated to a pair of planes, and the specific relation
between the shear and the lattice was then given by the orientation
of these planes. This was a very nice application of affine strain geometry, and for students there is no better example in the entire literature of solid mechanics. Developing this technique further, Taylor
and Elam proceeded to investigate *double slip* very thoroughly over
a period of years (during which nothing comparable was done elsewhere). At the outset (1923a) they conjectured that a second system
would be activated at the stage when its Schmid stress attained the
current value of the Schmid stress on the primary system (this was
in line with Taylor's mathematical idealization that all systems har-
dened equally, whether active or not). Taylor and Elam's collateral
experiments, however, did not substantiate the conjecture: they
found, in fact, that the onset of double slip was generally delayed
beyond the predicted stage (and often well beyond it). The over-
shoot was confirmed by further tests on aluminium crystals under
uniaxial tension (1925b) and later under uniaxial compression also
(1927b, 1927c). It is now known that the phenomenon occurs
much more widely, especially in other face-centred cubic crystals
and under a variety of loadings. The pioneering work of Taylor
and Elam on this topic still makes fascinating reading in its own

right; an authoritative commentary from a modern perspective has appeared in a monograph by K.S. Havner.[3]

Having gained this much insight into crystal mechanics, it was natural for Taylor to attempt a capital theoretical problem: to predict the bulk plastic behaviour of a random aggregate of crystals, each of which conforms to (i), (ii), (iii) and a generalization of (iv). Here he introduced a pragmatic simplification at the outset (1938a); namely, that every constituent crystal is homogeneously deformed and in the self-same way (not actually possible since the assemblage could not then be self-equilibrated). In general, five independent shears must be activated simultaneously to produce an imposed volume-preserving strain; Taylor proposed a heuristic principle (1938b) by which physically admissible sets of five could be selected from the large number of geometrically possible sets; he could then calculate the work needed to deform a crystal incrementally at any orientation in the aggregate, and finally average this work over all orientations. He performed the calculation for an aluminium aggregate under uniaxial load, and obtained encouraging correlation between the shearing stress–strain curve of the single crystal and the tensile stress–strain curve of the aggregate. Later, in 1951, the selection principle was shown by J.F.W. Bishop and R. Hill to follow rigorously from the properties (i) to (iv) above, and in the light of the other findings Taylor modified his original conclusions somewhat (1956a). Even now the problem still awaits a fully satisfactory solution, in which inhomogeneous deformation of the crystals would necessarily be admitted so as to equilibrate their interfacial tractions. Failing this, Taylor's pragmatic simplification is commonly used in, for instance, studies of the development of crystallographic texture in forming processes such as sheet rolling. Taylor himself was attracted briefly to this application (1927b) and, in the context of his compression tests on crystals in the form of thin disks, commented shrewdly on previous theories of texture.

Taylor terminated his work on single crystals in 1927 and thereafter turned to experiments on aggregates. The outcome was a famous memoir with H. Quinney on 'The plastic distortion of metals' (1931a). At

3. *Finite Plastic Deformation of Crystalline Solids*, chapter 1, Cambridge University Press, 1992.

that period the constitutive description of plastic response in an idealized isotropic continuum was being sought within a simple conceptual
framework, which had been codified over many years by Saint-
Venant, Lévy and von Mises, among others. Its essential elements
were (i) a yield function $f(\sigma_{ij})$ – namely some scalar invariant of the
Cauchy stress tensor σ_{ij} – whose critical magnitude marks a conven-
tionally defined limit of purely elastic response in a previously worked
metal; (ii) a plastic potential $g(\sigma_{ij})$ – another scalar invariant – whose
gradient $\partial g/\partial \sigma_{ij}$ at yield defines (except in magnitude) a potential tensor increment of plastic strain under the stress σ_{ij}; (iii) a hardening
rule, which specifies the dependence of the functional forms of f and
g on the prior cold-work. Taylor and Quinney concentrated on investigating (i) and (ii) for several metals in the form of thin-walled cylindrical tubes; just one type of pre-strain was applied, namely by
monotone uniaxial load, which was then combined in reduced amount
with an axial twisting moment. In previous experimentation of a somewhat similar kind, notably by Lode in 1926, the interpretation was
clouded by scatter due to unknown inhomogeneity and anisotropy in
the virgin tubes; by contrast, Taylor and Quinney's investigation was
distinguished by a well planned and largely successful procedure to
eradicate this source of scatter. For all metals tested they found that
the invariant f was not far different from $\sigma_{ij}\sigma_{ij} - \frac{1}{3}\sigma_{ii}\sigma_{jj}$, which was
the postulate of von Mises, and that the tensor $\partial g/\partial \sigma_{ij}$ deviated slightly
but consistently from $\sigma_{ij} - \frac{1}{3}\sigma_{kk}\delta_{ij}$ which was the postulate of Lévy
and von Mises. Many years afterwards, Taylor re-evaluated this 1931
data and found it to be in good agreement with what has become
known as the 'normality rule', $f \equiv g$ (1947a); in words, the plastic
strain-increment is co-directional with the local outward normal to
the current yield surface in a six-dimensional stress space.

To appreciate Taylor in the role of experimental physicist, *par excel-
lence*, one must study his papers with Farren (1925a) and Quinney
(1934a) in all their elaborate detail. In these he determined how the
macroscopic work expended in deforming a metal plastically is parti-
tioned between (a) heat to raise the temperature, and (b) strain-energy
stored at microscopic disarrangements of the atomic lattice. The
broad answer, for single crystals or for aggregates, was that (b) is of
the order of one tenth of (a), but the precise relationship is exceedingly

complex. Both experimental investigations were path-breaking in their contexts, especially the thermocouple technique in the first and the calorimetric technique in the second. That the outcomes were so successful is a remarkable tribute to the superb skills of Taylor's collaborators and to his own overall conception, which was allied to the most scrupulous assessment of every relevant factor.

Taylor made comparatively few incursions in the theoretical mechanics of solid continua, and did not join in its rapid development after World War II. Although he had a sound knowledge of the easier parts of classical elasticity, his mathematical equipment in relation to solid mechanics as a whole remained fairly circumscribed, more especially in regard to tensors (characteristically, he preferred the intuitive quadric representation to the formal transformation rule). Only once did he venture into deeper mathematical waters when, firmly supported by his student, A.E. Green (subsequently an elastician of international renown), he analysed fields of stress in perforated aeolotropic plates in connexion with failure modes in wood panels (1939a, 1945a). His principal solo contribution to elasticity was incidental to an attempt to explain the onset of bulk plastic deformation in metals whose flow stress falls at yield (1934b). To that end he proposed an elastic continuum whose shearing rigidity is suddenly lost within a spherical or elliptic cylindrical fault, so concentrating the stress locally and extending the yielded region still farther. The model itself did not win acceptance, but Taylor's analysis was a pioneering incursion into what is nowadays known as fracture mechanics (more precisely that part where classical elasticity is deployed at a microscopic or even atomic level).

In the continuum theory of plasticity at a macroscopic level, at a time when the foundations of the subject were still in contentious flux, Taylor had an unrivalled understanding of how the elementary constitutive principles should properly be framed (resting as they did on his own experiments between 1923 and 1931). Accordingly, when his help was sought on armaments design involving plastic deformation, he could give sound advice on theoretical problems beyond the capability of anyone else in this country in the early 1940s. One such problem had to do with the first-order deflexion of a pressurized diaphragm over an elliptical aperture (1942c); this was a relatively simple exercise for Taylor. His most brilliant achievement in the genre was a complete solution of the quasi-static radial expansion of a pinhole in a rigid-plastic sheet, where the chief unknown is the profile of the resultant coronet (found in 1943, published as 1948a); this was the first non-trivial plastic problem to be solved where the flow is non-steady and of unrestricted magnitude. Needless to say, he also conducted a typically ingenious experiment to check the theoretical predictions (but in this he was partly thwarted by cumulative errors in his numerical integrations). It is a matter for regret that Taylor did not go on to attack other plastic problems at this level of difficulty: for example, steady-state forming processes such as wire drawing, to complement his novel split-wire technique (1932a) for recording inhomogeneous internal deformations via scribed grids on the introduced interfaces.

Dislocations in metal crystals (by Professor Sir Nevill Mott)

Taylor's papers published in 1934 on the dislocation mechanism of plastic deformation in crystals had a very great influence on the development of the subject. It is true that in the same year Orowan and Polyani independently published two contributions ascribing plastic flow to the movement of dislocations; but what was unique about Taylor's work was a numerical expression for the work-hardening as a function of strain, containing a parameter, the drift distance of a dislocation before it was stopped, so that the stored energy could be calculated.

The main papers are: 'The mechanism of plastic deformation in crystals', (1934c, d), 'The strength of rock salt' (1934e) and 'The emission of the latent energy due to previous cold working when a metal is heated' (with H. Quinney, 1937a). In the first of these he begins with the experimental fact, established by himself and co-workers and others such as Schmid and Boas[4], that the stress–strain curve for many cubic crystals is parabolic, the stress S being related to the strain s by the relation $S \propto s^{\frac{1}{2}}$, the constant being temperature-dependent. He then goes on to describe rival theories as they existed at the time;

4. *Z. Phys.* **71**, 1931, p. 713.

some supposed that a perfect crystal was very strong, others that it was very weak. Taylor believed that the perfect lattice was very strong, but that what he called a dislocation could move for a negligibly small stress. This later became known as the 'Peierls force', and was shown to be not negligibly small in some materials. Implicitly he supposed that the generation of dislocations was easy; a full understanding of this had to await the discovery of the 'Frank–Read source' nearly two decades later. Then – and here was the key assumption – each dislocation could drift a distance L before being stopped, so that the strain produced by n dislocations per unit area (in Taylor's two-dimensional model) was $s = nLb$, where b is the magnitude of the Burgers vector, namely the distance which one plane slips over another when a dislocation passes.

The strain round a dislocation had been calculated in the first decade of the century by Volterra and by Timpe. It falls off inversely with the distance. Therefore, in a material with n stored dislocations per unit area, in which the average distance between dislocations would be $n^{-\frac{1}{2}}$, the stress necessary to move dislocations of opposite sign past one another would be of order $Gbn^{\frac{1}{2}}/2\pi$, where G is the shear modulus. Taylor identified this shear stress with the flow stress of the material, so that S must be proportional to $G(bs/L)^{\frac{1}{2}}$, giving the parabolic relationship. Taylor also considered the stress necessary to cause dislocations of opposite sign arranged regularly on a lattice to glide past one another, and he showed that the constant of proportionality in the above relation depended on the geometry of the dislocation arrangement.

Comparison with experiment is given in his papers for both metals and rock salt; Taylor deduced L from the observations, finding a value of order 10^{-4} cm, which he ascribed to the difference between surfaces of misfit. Also he showed that the density of dislocations deduced fitted reasonably with observations of the amount of stored energy.

Taylor's work was limited to cubic crystals; the parabolic law was not observed for hexagonal materials (e.g. zinc) where slip is only on one plane. His analysis supposed that the dislocations in a cold-worked material are uniformly distributed in space. I remember asking him once how he reconciled that with the observation of slip planes, and his answer was that he felt his paper to be a model and in no sense a final theory. More detailed theories certainly had to wait until three-dimensional descriptions of dislocations were given, the 'edge' and the 'screw', and dislocation arrangements were deduced from experiment. Taylor did not come back to the subject. His contributions, however, namely that the strength of a cold-worked material was due to the strains around dislocations, that the flow stress depended on dislocation density and arrangement, and that a stress–strain curve could be derived if a dislocation drift distance could be defined, were of major importance and showed subsequent workers that something quantitative could be attempted. Furthermore his calorimetric technique for measuring stored energy was the forerunner of many similar studies carried out subsequently by others.

Taylor's paper on dislocations was the beginning of an important branch of solid-state science, to which several textbooks have now been devoted, notably those by Nabarro, Cottrell and Friedel. His two major concepts have remained valid, namely that deformation of crystals is by movement of dislocations, and that what stops them from moving and gives strength to materials is internal strain. Among the interesting post-war developments were several ways of actually observing dislocations. An etch pit could be formed at the point where a dislocation ends at a surface, and dislocations could be 'decorated' and made visible under the optical microscope, notably in silver bromide crystals under illumination, precipitating silver along them. And in Cambridge Hirsch showed that in thin metal films they could be seen by transmission electron microscopy, appearing as little lines moving across the screen. I vividly remember the day in the Cavendish Lab. when my young men burst into my room to say that they had seen a moving dislocation.

Taylor's 'Note on the early stages of dislocation theory'

In 1963 G.I. was invited by Dr C.T. Smith, organizer of the Sorby Centennial Symposium on the History of Metallurgy, to record his memory of the development of his ideas about dislocations. He did so, after refreshing his memory by correspondence with A.A. Griffith, and here is the note he prepared for publication (1965b). According to G.I.'s accompanying letter to Dr Smith, the 'big leap forward' came when he gave up trying to analyse the stress in the neighbourhood of a dislocation in terms of elliptic co-ordinates and instead thought of dislocations as singularities with known associated stress distributions which could be superposed. G.I. said 'the rest follows simply', but it seems to me he may here be making the concept of a dislocation seem to be purely mathematical for the purposes of a reply to Dr Smith's enquiry.

> One of the problems presented to me at the Royal Aircraft Factory in August 1914 was that of calculating the strength of aircraft spars in bending and torsion. To solve such problems, A.A. Griffith and I developed the soap-film analogy. We found that keyed crank-shafts of propellors with only slightly-rounded internal corners often did not fail, even though the calculated stress concentration was still large, and we inferred that the material must be behaving plastically.
>
> In 1915 I left Farnborough to join the Royal Flying Corps, but Griffith carried on with this work. The fact that metals used in engineering do not apparently suffer the high stress concentration predicted by elastic theory led him to experiment with brittle materials which may break through rapid extension of a crack. If the crack is considered as truly sharp-ended and the material is treated as continuous and elastic, the stress concentration would be infinite and the material would be incapable of resisting even the smallest stress if it had a crack in it. This paradox was resolved by Griffith who showed, using Inglis' equations for the stress round an elliptic hole, how the elastic theory can be used to calculate, not the stress concentration, but the reduction in elastic energy due to extension of a crack in a stressed elastic material. This is finite even though the calculated stress is infinite at the tip of the crack.
>
> By equating this loss in elastic energy to the gain in surface energy when the crack extends, Griffith obtained an expression for the strength of a material containing cracks. Before Griffith's theory calculations had been made in which the stress required to separate two surfaces was calculated by equating the surface energy to the work done against this stress when two surfaces are separated through the short distance at which attractive molecular (Van der Waals) forces extending across the gap have become weak. This method predicts strengths which are far greater than those actually observed. Griffith's theory by contrast

was to my mind the first real advance in understanding the strength of materials.

My interest in solid state physics lay dormant till 1922 when I was present at the meeting of the Royal Society at which Professor H.C.H. Carpenter and Miss Constance Elam exhibited large crystals of aluminium which had been deformed plastically in a way which obviously had some connection with their crystal structure. This led to my collaboration with Miss Elam (who later became Mrs Tipper) in which we investigated this question and determined macroscopically the modes of plastic distortion of single crystals of several metals under various conditions of straining. In the course of this work we frequently had occasion to measure the stress–strain curves and in most cases these were nearly parabolic, the stress necessary to increase strain increasing with the amount of strain. In the Griffith crack theory the enlargement of a single crack reduces the stress necessary for further extension so that it cannot be used directly to form a model of a strain-hardening material.

During the years 1930–34 I had, in collaboration with Mr H. Quinney, been measuring the difference between the amount of energy used in cold work when straining a plastic material, and the amount of elastic energy locked up in possible self-strained faults, and so to get another measurable quantity to link up with theoretical models of the structure of a strain-hardened material. These studies did not help in forming a quantitative theory of metal structure but they did confirm the idea that increase in the elastic energy of self-strained internal faults was associated with increase in resistance to strain.

During the years 1928–33 I was more actively engaged in hydrodynamic research, but retained an interest in solid state physics and read some of the theories put forward to account for the strength of metals. Most of these seemed to be incapable of being expressed in quantitative form and it seemed to me that unless this could be done a theory had little value. One of the most popular was that there exist slip planes in metals which can slide with perfect ease, but that when they do portions of the material come loose and get turned round and form keys which hold up the slip. This theory seemed to me very vague and certainly untrue in the case of the face-centred cubic metals with which Miss Elam and I had been dealing, because we had shown that any assigned deformation would occur by an appropriate combination of slipping on similar crystal planes so that the 'key' would deform with the rest of the material even if the material in it was at a different crystal orientation from that of its surroundings.

Among the experimental researches I was particularly interested in the experiments published during those years in Italy, Russia and Germany, on the plasticity of rock salt, because its stress–strain characteristics were like those of metals and yet there was evidence that its strength was sensitive to cracks or scratches like Griffith's brittle materials. While trying to understand how the existence of cracks would produce the effect of strain-hardening, I expressed the view, at a meeting of the Faraday Society in 1928, that macroscopic strain must correspond with an increasing number rather than increasing length of crack and that this must lead to an increase in the ratio of mean stress to the maximum stress at the end of the cracks. An increase in the number of cracks could be due to long cracks, particularly cracks on shear lines, sealing up again in the middle and thus producing two cracks out of one. I did, however, notice one significant difference between the two cracks produced in this way and the original crack, namely that one of them would have more material above it than below while the reverse applied to the other.

Though this model of plastic strain was partially satisfactory from a qualitative point of view, I did not see how to translate it into quantitative form because I was thinking in terms of the analysis of Griffith and Inglis, which depended essentially on the use of elliptic co-ordinates. These co-ordinates, in which the variables are constant over a series of confocal ellipses and hyperbolas, are suitable for discussing the stress distribution round a single elliptic hole and can therefore represent the crack. It would be practically impossible to use elliptic co-ordinates to discuss the stress distribution when there are two or more cracks in the field.

The critical thought which made possible further development of these ideas came in 1933 when I realised that if the material on the two sides of a Griffith shear crack joined together again after the passage of its ends the stress round each end would become independent of that at the other as they separated and could therefore exist in isolation. This reduced the problem to the consideration of stress systems in the neighbourhood of what the mathematicians call singularities and had the great advantage that the stress systems of any number of these singularities could be superposed.

As soon as I realised this I looked up Love's textbook on

elasticity to find out whether the stress distributions corresponding with such singularities, or dislocations, had been calculated. I found that Timpe had given the particular solution I required in 1905 and that the matter had been followed up and generalised to other types of dislocation by Volterra in 1907.

Having found a method for calculating the internal distribution of stress in a material containing dislocations it was an obvious step to assume that the dislocation was formed by transfer of atoms over one atomic distance and thus to fix the distribution of elastic stress round the singularity. It was also a simple matter to connect the relationship between the overall strain and the number of dislocations in a simple two-dimensional model. To complete the model the reaction of one dislocation on another was assumed to operate in such a way that the atoms would only jump across a dislocation when the combined action of externally applied stress and that due to all the rest of the dislocations rose to a certain critical value. When I found that this model led to a parabolic relationship between stress and strain and that the spacing of the dislocations which it entailed was physically reasonable I was naturally pleased, but I never regarded the two-dimensional model described in my papers of 1934 as anything but a suggestive model. In 1934 I was developing the statistical theory of turbulence and experimenting with anchors. These activities took up most of my time so I did not attempt to develop the work on dislocations beyond the state described in the two papers published that year (1934c, d).

CHAPTER 12

Turbulence: a challenge

Horace Lamb has a section on turbulent motion of fluid in his classic book *Hydrodynamics*, and it begins with the remark, 'It remains to call attention to the chief outstanding difficulty of our subject'. This sentence appeared first in the second edition published in 1895, and it remained in the sixth edition published in 1932. And if Lamb lived today and contemplated a seventh edition I am sure he would see some justification for regarding the remark as still appropriate. Turbulent flow[1] was and is both the most difficult, and, owing to its common occurrence in nature and in engineering contexts, the most relevant of the unresolved areas of fluid mechanics. The importance of turbulent flow has been widely recognized for many years, but attempts to derive fundamental relations between the velocity fluctuations characteristic of the turbulence and quantities of practical interest – such as rate of dissipation, rates of transfer of conserved quantities, dispersion, and forces exerted on rigid boundaries – have been largely unsuccessful. However, much useful progress was made in the period between the wars, and G.I. was at the centre of it. We take first the period prior to 1935, because there was a marked change in the approach to the subject (introduced by G.I.) in that year.

In one of his few review-type articles G.I. described the current state of turbulence research in chapter 5 of the well-known book *Modern Developments in Fluid Dynamics*.[2] It is concerned to a large extent with G.I.'s own work, and so may be useful to readers as a more

---

1. Which we can define descriptively as flow in which the velocity of the fluid is a random function of position and time and inertia forces on the fluid are significant.
2. Edited by S. Goldstein, Oxford University Press, 1938.

detailed alternative to the following account. Much later G.I. published two short and readable accounts of his early ideas presented at international symposia, the first being about turbulent diffusion (1959f) and the second about turbulence generally (1970b).

G.I.'s essay for the 1915 Adams Prize

G.I.'s first work on turbulence as Schuster Reader in Dynamical Meteorology in 1912 was to make simple observations of fluctuations in the direction and magnitude of the wind velocity at points within a few metres above flat ground, as already described in chapter 4. For the further fruits of his work during the period 1912–14, which includes six months at sea on the s.s. *Scotia*, we may refer to the essay that he submitted for the 1915 Adams Prize at Cambridge. This is a valuable prize with high prestige which is awarded for the best account of original research in a field of physical science specified every two years. G.I.'s essay was prepared hurriedly, because as a consequence of his urgent work on army aeroplanes at the Royal Aircraft Factory he had very little free time during the five months prior to the latest date for submission (31 December 1914). In his preface G.I. says, disarmingly, that 'the diagrams were drawn and the formulae filled in by my Father and Mother, without whose help I should have been unable to get any of the papers into a form suitable for sending in'. Although the essay shows signs of hasty preparation (and there are now some missing minor parts) it is remarkable for the originality and depth of the research described there and it is not surprising that the prize was awarded to G.I.

The specified field for essays submitted for the 1915 Adams Prize was 'The phenomena of the disturbed motion of fluids, including the resistances encountered by bodies moving through them'. G.I.'s essay was entitled *Turbulent Motion in Fluids* (although the contents were broader than is suggested by that title). The rather whimsical regulations at that time required a motto on the cover of his essay to be the means of its identification instead of the author's name, and G.I. chose for his motto the very appropriate couplet:

I saw you toss the kites on high
And blow the birds about the sky.[3]

The essay consists firstly of his big paper (1915a) describing observations of the vertical distributions of temperature, water-vapour content and wind velocity in the friction layer of the atmosphere and their comparison with theory based on the assumed existence of the eddy transfer coefficients suggested some years earlier by Boussinesq. This systematic use of eddy transfer coefficients for the analysis of a comprehensive set of data was a major development. As already noted in chapter 5, the paper (1915a) also includes a deduction of the spiral distribution of wind velocity in the friction layer of the atmosphere due to Coriolis forces; and there is a discussion of the influence of a rigid boundary on the stability of unidirectional flow when the fluid has very small viscosity. Secondly, there are three chapters of the essay on the calculation of various turbulent flow fields on the assumption that mean stresses in the fluid may be represented in terms of an eddy viscosity. Thirdly, there are chapters on steady unidirectional flow of stratified fluid in which criteria for stability are found in the form of inequalities involving the undisturbed vertical density and velocity gradients in a dimensionless ratio now termed the Richardson number. This work on stability of stratified fluid flow was an early version of that described in a well-known later publication by G.I. (1931b), as mentioned in chapter 9. Fourthly, the last two chapters describe the work on motion of a solid body in fluid which is rotating as a whole. This last investigation has already been referred to in the context of G.I.'s research soon after the end of World War I (see chapter 7). The various parts of the essay do not hang together well, but it is a masterly account of research on some very difficult problems in fluid mechanics, turbulent flow in particular.

3. These are the first two lines of the poem entitled 'The Wind', from *A Child's Garden of Verses*, by Robert Louis Stevenson. A copy of this book was awarded to G.I. as the Smith's Prize for 1910, for his work on shock waves (see chapter 4). He may have made this strangely juvenile choice of the award in recognition of a family connection with the illustrator, Millicent Sowerby.

The semi-empirical theories of turbulent transfer

In the twenties and early thirties the main research topic associated with turbulence was the mechanism of transfer of conserved quantities across approximately parallel streamlines of a steady mean flow. The issues under discussion in what were referred to as the 'semi-empirical theories' were as follows:

(a) What are the properties and significance of an eddy transfer coefficient ϵ such that the mean rate of transfer of any conserved scalar quantity with intensity θ in the direction y in which the mean flow velocity U varies is $\epsilon d \langle \theta \rangle / dy$?

(b) What is the corresponding expression for the rate of transfer of momentum (that is, the Reynolds stress)?

(c) Can it be assumed that the transfer of any conserved quantity θ proceeds by convection of an element of fluid without change of θ over a distance l in the y-direction followed by mixing of the element with its new environment, whence, as in the kinetic theory of gases, $\epsilon = lv$, where v is a statistical measure of the y-component of the fluid velocity?[4]

In his pioneering paper 'Eddy motion in the atmosphere' (1915a) G.I. had already made significant progress with these issues. He answered (a) empirically from observations of vertical distributions of temperature and water-vapour content in the air over the ocean. In connection with (b) and (c) he recognized that fluid momentum is not a conserved quantity in lateral movements of the fluid, because in general fluid momentum is affected by pressure gradients; and so the idea of a mixing length for momentum may not be correct. In these circumstances it seemed to him to be worth noting that in a wholly two-dimensional turbulent motion in the (x,y)-plane the

---

4. It seems evident physically that the 'mixing length' l should be chosen to be comparable with the length scale of the energy-containing eddies. However, there is then the difficulty that in practice the latter length scale is seldom small compared with the distance over which the gradient of $\langle \theta \rangle$ is constant, which conflicts with the above (implicit) hypothesis that the eddy transfer coefficient is a local quantity. There is thus a lack of self-consistency in the transfer model implied by the above questions, strictly speaking.

vorticity associated with a fluid element is constant (for an inviscid fluid) and so *is* a conserved quantity in the above sense, in which event the rate at which the mean momentum of unit volume of the fluid increases may be shown to be of the form

$$M' = \rho\epsilon\frac{d^2U}{dy^2}.$$

Working in ignorance of these findings by G.I. 10 years later, Prandtl[5] assumed that momentum could be regarded as a conserved quantity, in which event the rate of increase of mean momentum is of the form

$$M'' = \frac{d}{dy}\left(\rho\epsilon\frac{dU}{dy}\right).$$

There was much discussion of the applicability of these two expressions, and of their consequences for the properties of various cases of turbulent shear flow with approximately parallel mean-flow streamlines. G.I. made what he called a searching test of the comparative merits of the two 'theories' by comparing the distributions of mean temperature and velocity in the wake behind a heated cylinder (1932c). He found results more in accordance with the predictions of the vorticity transfer theory than with those of the momentum transfer theory. However, other similar tests on wall-bounded turbulent flows later showed the reverse (1935g, 1937c).

The dependence of the mixing length *l* on *y* needs to be known before either of the above two expressions can usefully be substituted into the equation of motion, and here too there was much debate in the twenties. Von Kármán joined in by proposing an expression for *l* which depends only on the local properties of the mean flow, viz.

$$l = \frac{dU}{dy}\bigg/\frac{d^2U}{dy^2},$$

in contrast to the more common expressions for *l* which depend on the flow geometry; and by combining this with Prandtl's hypothesis

5. *Z. Angew. Math. Mech.* **5**, 1925, p. 137.

that near a plane wall l is proportional to the distance from the wall, one finds the famous 'law of the wall', viz. $U \propto \log y$.[6]

G.I. never regarded the so-called vorticity and momentum transfer theories as rivals, and I believe he was content to regard the vorticity transfer theory simply as a note about turbulent transfer when the flow is approximately two-dimensional, as appeared to be true in the case of the wake behind a cylinder. Looking back, in 1970b, he wrote: 'I was not satisfied with the mixture-length theory, because the idea that a fluid mass would go a certain distance unchanged and then deliver up its transferable property, and become identical with the mean condition at that point, is not a realistic picture of a physical process.' Interest in the semi-empirical theories of eddy transfer began to wane in the mid-thirties, and they have now been largely superceded by more sophisticated turbulence 'modelling' backed up by powerful computing facilities. Nevertheless data obtained to test the theories is still valuable; see for example G.I.'s paper on the distribution of mean velocity and temperature in the turbulent flow between two concentric cylinders with the inner one rotating and being heated (1935g).

The statistical description of turbulence

G.I.'s greatest contributions to our understanding of the turbulent motion of fluids were made in the mid-thirties. The connections with his own work 15 and more years earlier are not well-known, and are often overlooked, but in fact they are close, especially with the papers (1915a), (1917d) and (1921b), and show that this tremendous burst of progress was not the result of a sudden inspiration but was the outcome of deep and systematic thought over 20 years.

We should note in particular G.I.'s seminal paper 'Diffusion by continuous movements' (1921b). This short and simple discussion of probability theory for random functions led to a novel mathematical investigation of the statistical relation between the velocity and the displacement of an element of fluid in turbulent flow. The move-

6. First announced by von Kármán at the 3rd International Congress of Applied Mechanics in Stockholm in 1930.

ments that lead to transport and diffusion in turbulent motion of a fluid are continuous, in contrast to the motions of molecules in a gas, and G.I. realized that a mathematical study of the properties of continuous random functions would be needed before turbulent diffusion could be described quantitatively. He made the first steps in such a study, and perceived the simplifications that are possible when the velocity of a material fluid element is statistically stationary. The main result of the paper is the now classical formula

$$\frac{d\langle X^2 \rangle}{dt} = 2\langle u^2 \rangle \int_0^t R(\xi)d\xi$$

relating the rate of increase of the mean square of the displacement of the fluid element in one direction to the integral of the coefficient of correlation of the velocity component u in that same direction at two different times, given that u is a stationary random function of t. Taylor noted that since $R(\xi)$ is unity when $\xi = 0$ and could be expected to fall rapidly to zero as $\xi \to \infty$, this relation predicts simple general features of the lateral spread of a (cold) smoke plume from a chimney in a horizontal wind. The distance downstream over which the plume is carried by the mean wind is a measure of time here, and so we find conical lateral spreading near the source (where $t \to 0$) and a paraboloidal shape far downstream (where $t \to \infty$), both these features being in accordance with observations.

The above formula shows that, when t is large and the integral is constant, the diffusion of marked elements of the fluid can be represented in terms of a constant eddy transfer coefficient equal to

$$\langle u^2 \rangle \int_0^\infty R(\xi)d\xi.$$

We see that the integral can then be regarded as a 'mixing time', analogous to, but more precise than, the commonly used mixing length. Note that these developments rest on the assumptions that the statistical properties of u are independent of t and that the mean velocity of the fluid is uniform (and so plays no part in the problem).

G.I. perceived that making use of the velocity correlation function and similar statistical quantities would require some unfamiliar mathematical techniques, and he said in the paper, which was published by the London Mathematical Society: 'Such questions might be examined with advantage by a pure mathematician.' The paper was well ahead of its time, and appears to have been the first to introduce velocity correlation functions into the study of turbulence. It was not until 14 years later that laboratory experiments on diffusion in turbulent flow were related to this simple theory and that the idea of using probability theory in the description of turbulent velocity fluctuations was adopted whole-heartedly. And it was G.I. himself who made these further developments.[7] Meanwhile it was an early notification that mixing-length theories which appealed to discontinuous processes like those occurring in the kinetic theory of gases were not needed. Random fluid movements are normally continuous, and G.I. had shown that there exists an appropriate means of representing continuous movements.

The novel contribution of 'Diffusion by continuous movements' was to show how to describe diffusion in terms of the statistical properties of the fluid velocity defined in the Lagrangian manner as the velocity of a material element of fluid, and so as a (continuous) function of time. The new way of thinking about turbulent motion of fluid put forward by G.I. in the mid-thirties on the other hand was characterized by the representation of fluid velocities defined in the Eulerian way as continuous random functions of time and position in space to be described statistically. The advantage of a description of the flow field in terms of fluid velocities defined in the Eulerian, rather than the Lagrangian, way is that it simplifies greatly the expressions for the instantaneous spatial derivatives of velocity that occur in the equation of motion (and in the vorticity and the rate of dissipation).

The message now seems to be quite obvious, but there were no antecedents except G.I.'s own 1921 paper, which had attracted very little attention among applied mathematicians, and a little-known mathematical investigation of velocity correlation functions by

7. In the different world of probability theory, Norbert Wiener said in his autobiography that his method of handling random functions published in 1930 was directly developed from Taylor's 'theory of turbulence'.

Keller and Friedman[8] in 1924. G.I. took the investigation of turbulence down to a deeper level and revealed the language and the concepts that future research must use. The implications were not grasped quickly and for a couple of years after 1935 he had the field to himself. The merit of the dozen or so papers in the period 1935–38 in which he showed how to describe turbulence in statistical terms and how to use that description, for example in a derivation of the consequences of the equation of motion, lies in their advocacy of a quite novel approach, rather than in the discovery of new results. They contain in fact rather few specific results (a reflection of the subject – we still lack specific results about turbulence). The approach has now been almost exclusively adopted, and greatly extended, and I think present-day students seldom have need to go back to these papers.

There were two coincident developments which encouraged G.I. to bring to publication his four connected papers entitled 'Statistical theory of turbulence' (1935c,d,e,f). No way of recording local velocity fluctuations accurately existed in the twenties, and G.I. saw little point in constructing theories which could not be connected with measurable quantities. The position changed in the early thirties, when the hot-wire anemometer (a fine heated wire whose temperature fluctuates with the velocity of the cooling fluid in which it is immersed) was developed as an instrument which allowed rapid fluctuations in velocity of the fluid to be measured and analysed electronically. For experimental work on turbulence in wind tunnels using the new anemometers G.I. enlisted the help of L.F.G. Simmons at the National Physical Laboratory (at Teddington, near London) since he was not skilled in electronics.

G.I.'s active work on turbulence in the thirties was also aided by his increasing acquaintance with turbulent flow systems relevant to aerodynamics and wind tunnel measurements, which supplemented his early experience derived mainly from meteorology. Sir Arnold Hall, who was a research student in the Department of Engineering at Cambridge working under G.I.'s supervision during the late thirties,

8. *Proc. 1st Intern. Cong. Appl. Mech.*, 1924, p. 395.

expressed the view that G.I.'s research on turbulence 'perhaps owed a little inspiration to his close association with the work B.M. Jones was doing simultaneously in the Aeronautical Laboratory at Cambridge. Jones was concerned with the drag of bodies in a fluid stream, particularly with the skin friction resistance of aeroplane wings, and was conducting experimental work both in the laboratory and in flight. This led him to a study of boundary layers in their laminar and turbulent forms. Taylor took a close interest in this work.' This was no doubt the origin of G.I.'s interesting attempt to relate the level of turbulence in a stream with the position of transition from laminar to turbulent flow in the boundary layer on a sphere and hence to the drag on the sphere (1936b).

One of the important by-products of these papers was the demonstration of the value of theoretical and experimental studies of the decaying turbulence generated by a grid of rods placed across a wind-tunnel stream. The correlation of the velocity fluctuations at two different points is a function only of the spatial vector joining the two points when the turbulence is spatially homogeneous, and G.I. recognized that the turbulence behind a grid of rods would have approximately this property, except near the grid. He noted also that the grid does not impose any strong directional preferences and that the turbulence generated by it is approximately isotropic. This was a great simplification, and enabled many of the theoretical connections between different statistical quantities, such as those that follow from the mass-conservation relation, to be confirmed by measurement. It was not quite such a simple state of turbulence as G.I. initially supposed, and he mistakenly assumed that the mean-square velocity, but not the characteristic length scale of the turbulence, would change with distance downstream from the grid. Measurements show that the kinetic energy of the turbulent motion per unit volume decreases as the fluid is carried downstream and that the length scale increases. We still have no simple theory which accounts quantitatively for these decay relations, but grid turbulence has undoubtedly been the main testing ground for theoretical relations.

A festschrift in *ZAMM* was organized in 1934 to commemorate Prandtl's 60th birthday, and G.I. contributed a paper of lasting significance (1935b). Some years earlier Prandtl had shown by simple arguments that the longitudinal and transverse components of velocity due to a steady (that is, one which is independent of longitudinal position x) disturbance upstream of a wind-tunnel contraction would change on passage through the contraction by multiplicative factors C^{-1} (from Bernoulli's theorem) and $C^{\frac{1}{2}}$ (from a consideration of extension of vortex lines) respectively, where C is the ratio of the upstream cross-sectional area to the downstream area. This information was useful for the design of wind tunnels. However, the utility was greatly enhanced when G.I. pointed out that, if the contraction ratio C is so large that the relative velocity of two neighbouring elements of fluid due to the contraction is much larger than that due to the disturbance, then the latter contribution may be ignored; and a means of investigating the effect of the contraction on a general turbulent stream, steady or unsteady, is thus available. The essence of the method is to suppose that the effect of the rapid contraction is approximately to redistribute the vorticity of the disturbance by a uniform extension C in the stream direction and uniform contraction $C^{-\frac{1}{2}}$ in each of two orthogonal directions in the transverse plane in the case of a wind tunnel of circular cross-section. It seemed not to be possible to obtain an explicit general solution of the resulting equations for the disturbance, but G.I. solved them for the case of a disturbance which is sinusoidal in each of three orthogonal directions and explored some special cases, including those for which the disturbance is steady.

This was the beginning of what has become known as Rapid Distortion Theory. As with so many of G.I.'s papers, an idea which appears at first sight to be special, and associated with a particular context, proves to be of wide applicability. RDT has now been applied to a number of situations in which a large-scale deformation is suddenly imposed on turbulent flow.

Another innovation to be noted in this festschrift paper by G.I. was his use of Fourier analysis of the spatial distribution of the velocity fluctuation. With the aid of RDT he had been able to calculate the effect of a wind tunnel contraction on a stream in which a (small) fluctuation in fluid velocity varies sinusoidally in all three directions (1935b). The investigation was successful, as far as it went, but G.I.'s real objective was the effect of the contraction on turbulent velocity fluctuations and he was held up by not knowing how to 'integrate' the results over all the Fourier components of a turbulent, and hence random, fluid velocity. He made a rather similar use of a sinusoidal velocity distribution a little later in a short investigation of the relation between the mean-square fluctuations in pressure, pressure gradient and velocity, and again was obliged to regard a single Fourier component simply as a typical form of velocity fluctuation (1936a).

The same question concerning the significance of a flow field in which the fluid velocity varies sinusoidally with position arose in a paper with Albert Green about the way in which energy is transferred from large eddies to smaller ones in a homogeneous turbulent flow (1937b). G.I. had for many years seen this as an important fundamental process in turbulence, and he proposed here to investigate it deterministically by taking as an initial condition the velocity distribution

$$u = A \cos \alpha x \sin \beta y \sin \gamma z, \quad v = B \sin \alpha x \cos \beta y \sin \gamma z,$$
$$w = C \sin \alpha x \sin \beta y \cos \gamma z,$$

which satisfies conservation of mass if

$$\alpha A + \beta B + \gamma C = 0.$$

According to the Navier–Stokes equation, the expression for the velocity at subsequent times contains discrete higher harmonics (the 'smaller eddies'), and Taylor and Green calculated all the harmonics in a power series in t up to the term t^3 for general values of $A, B, C, \alpha, \beta, \gamma$ and up to t^6 for the special choice $A = -B$, $C = 0$, $\alpha = \beta = \gamma$. The results showed, as expected, that the mean-square vorticity decreased with time t when the Reynolds number was small but *increased* when the Reynolds number was large. There was an indication that the mean-square vorticity reached a maximum and began to decrease at later times at these higher Reynolds numbers (as would be expected), but the power series was then inaccurate. The further time development of the Taylor–Green flow has been studied intensively with the aid of computers. This kind of specific calculation appealed to G.I. as a means of demonstrating the mechanical processes at work in turbulent motion.

The final and crucial step in elucidation of the role of Fourier analysis was taken in G.I.'s paper (1938d). Here he showed by rather heuristic but adequate mathematics that the Fourier transform of the two-point velocity correlation is the spectral density function representing the contribution to the mean-square velocity from the different harmonic components.[9] This was a real advance, because mathematical results which had been obtained for a sinusoidal velocity distribution could now be converted to more significant statements about the energy spectrum of the turbulence. Moreover it allowed the velocity correlation function to be obtained from electronic analysis of the signal from a hot-wire anemometer. Simmons at the NPL had made measurements for G.I. of both the two-point velocity correlation and the frequency spectrum of the velocity at a point, in the turbulence generated by a grid of bars in a wind tunnel. G.I. showed that they satisfied the Fourier transform relation, as would be expected if the spatial distribution of fluid velocity changes relatively slowly as it is carried past the point at which the frequency spectrum is measured (the frozen-pattern hypothesis – another much-used Taylor idea).

Simmons's measurements of the spectrum at different tunnel speeds all had the same shape, except at the highest frequencies where the amount of kinetic energy increased with the stream speed. In the second-last sentence of the paper G.I. wrote: 'the fact that small quantities of very high frequency disturbances appear, and increase as the speed increases, seems to confirm the view frequently put forward by the author that the dissipation of energy is due chiefly to the formation of very small regions where the vorticity is very high.' Strictly speaking, all that 'the fact' confirms is that the non-dimensional ratio of mean-square vorticity to mean-square velocity increases with the Reynolds number, but the additional intuitive inference of spottiness of the vorticity distribution is absolutely correct, as research after the war was to show. This prediction of spottiness of the vorticity distribution was first made by G.I. in (1917d), as we noted in chapter 4.

9. G.I. attributed the general mathematical form of this theorem to Norbert Wiener, *The Fourier Integral*, Cambridge University Press, 1933.

G.I.'s clear view of the extension of vortex lines as the way in which the mean-square vorticity is built up to a very high level in a turbulent motion was set out in two papers (1937d, 1938c) apparently written mainly to correct a misunderstanding in some work done by von Kármán[10] after seeing the initial set of four papers by G.I. Von Kármán had improved G.I.'s rather primitive methods of working out the analytical consequences of statistical isotropy, and he was able to obtain the formal expressions for the rate of change of the velocity correlation and the rate of change of mean-square vorticity from the Navier–Stokes equation. The rate of change of mean-square vorticity is the difference between two terms, one a mean product of three velocities or their derivatives and the other a mean-square vorticity gradient representing the rate of decay due to viscosity, and in the interests of tractability von Kármán assumed the triple correlation term could be neglected. G.I. politely but firmly pointed out that the term neglected by von Kármán represents the rate of production of vorticity by the extension of vortex lines, a fundamental process in turbulent motion which is responsible for the high rate of dissipation; and he showed from some measurements of grid turbulence made by Simmons that in this particular case the term neglected by von Kármán was three times as large as the rate of decrease of mean-square vorticity. G.I. had had this understanding of the role of inertia forces in the mechanism of turbulent motion for more than 20 years, but von Kármán's error brought a clearer explanation from him than he had previously given.

Enter Kolmogorov

'The spectrum of turbulence' (1938d) was G.I.'s last paper on turbulence before the needs of World War II took him away from academic research. It is interesting to speculate on what further aspects of turbulence he would have chosen to investigate had the war not intervened. I personally had good reason to speculate, because when I arrived in Cambridge in April 1945 to work for a PhD on turbulence under his supervision I found to my surprise that he was too absorbed in

10. *J. Aero.* **4**, 1937, p. 131.

research arising out of defence problems to wish to take up work on turbulence again. He had no specific plans for the research that I should undertake, and, while always being interested in my progress was content to let me do what I wished. At the beginning I cast about for new ideas, and in the course of searching the literature came across two brief papers published in the USSR in 1941 in which Kolmogorov put forward the idea of statistical equilibrium of the small-scale components of turbulent motions and which had miraculously made their way safely to a Cambridge library. I was excited by them, and told G.I. what I had found.

A little later, during the summer of 1945, G.I. was asked if he would allow a visit by W. Heisenberg and C.F. von Weizsäcker, who were being held in a large house not far from Cambridge together with eight other German atomic scientists and who wished to discuss some new work on turbulence with him. Having lots of time on their hands but no facilities, the German scientists had chosen to think about difficult problems which were easily formulated, and the experience of lying on the grass and watching the clouds revived an interest in turbulence which Heisenberg had had since the time of his doctoral dissertation under Sommerfeld. Heisenberg and von Weizsäcker described to G.I. some new ideas on the spectral representation of isotropic turbulence and the transfer of energy from large to small-scale components, and although these ideas were embedded in unfamiliar analysis it was evident to G.I. and me that they had points in common with Kolmogorov's theory. To add to the coincidence, it was later noticed that in 1945 the distinguished physical chemist L. Onsager at Yale University had published an abstract in which essentially the same idea of statistical equilibrium of the small-scale components was put forward independently.[11] I described these three independent developments in a paper presented at the 6th International Congress for Applied Mechanics in Paris in September 1946, and the text was later published in *Nature* (**158**, 1946, 883–4).

On looking back at this incident I recall that G.I. did not draw attention, although he would have been entitled to do so, to the close-

11. *Phys. Rev.* **68**, 1945, p. 286.

ness of both Kolmogorov's theory and these ideas of Heisenberg and von Weizsäcker to some of his own much earlier work. The concepts of an energy cascade of increasing wave-number extent as the Reynolds number is increased, and of confinement of the effect of viscosity to the largest wave-numbers, and of some kind of statistical universality of the small-scale components, and of intermittency, were all familiar to him, as is indicated by his paper on atmospheric turbulence (1917d). The important idea he appears to have missed was that the statistical quasi-equilibrium of the small-scale motions depends on such a small number of parameters, namely two, the rate of energy dissipation and the fluid viscosity, that dimensional arguments alone yield explicit results.

This is a point of some historical interest, so it is worthwhile to see what G.I. himself said about the common ground of his early work and the new developments by Kolmogorov and by Heisenberg and von Weizäcker and by Onsager. The issue was examined in my article 'An unfinished dialogue with G.I. Taylor'[12], from which the following is an extract:

> **G.K.B.** Can you recall whether the theory that is now associated with Kolmogorov's name was in fact known to you before 1941, implicitly if not in the form in which he put it forward? And if so, can you remember the development of your ideas about the small-scale components of turbulent flow from 1915 on?
> **G.I.T.** Yes, I did have ideas about the eddy cascade in 1917 but did not see how to express them in mathematical form. The attempt I made to do this starting with a particular form of eddy is described in my paper with A.E. Green in 1937, but it was not successful and I certainly did not get on to the kind of statistical similarity argument which Kolmogorov, Heisenberg and Onsager developed independently of one another a few years later. As you say, I did realize as far back as 1917 that there must be something that gives small-scale turbulence a statistically isotropic character and that this would be a result of some universal quality in the grinding-down process. I could see that for separations of pairs of points outside the range in which velocity gradients can be regarded as constant there should be a range in which a universal law of grinding down determines the

12. *J. Fluid Mech.* **70**, 1975, pp. 625–638.

relationship between the correlations and the separation of two points in a turbulent field. However I did not see how to turn this idea into a mathematical description which could form the basis of a theory and could predict things that could be verified or disproved experimentally.

Before 1935 I did not see how statistical definitions of Eulerian correlations could be used for any other purpose than as one element in the mere kinematic description of a turbulent field, and I did not want to publish anything until I found something that could be verified experimentally.

G.K.B. I suppose the step which made possible the many useful deductions from the similarity theory by dimensional arguments alone was the recognition that the rate of transfer of energy from one eddy size to the next smaller size is the only relevant physical parameter, aside from the fluid viscosity, in the determination of the structure of the small-scale components of turbulence. Had this idea occurred to you before we first came across Kolmogorov's work in 1945?

G.I.T. Though in 1937 I had realized the equivalence of the correlation description of turbulence and the spectrum description, my idea of the dynamics was directed to trying to connect the rate of increase of mean-square vorticity with dispersion, because if two neighbouring points on a vortex line are separating the vorticity is increasing, and of course the rate of dissipation of energy is increasing. This idea was expressed in a suitable mathematical form by Kármán & Howarth for isotropic turbulence, but it did not lead to the law of grinding down in scale of eddies and I do not think I had any idea that a similarity argument could be used for this purpose till 1945 when you discovered those 1941 papers by Kolmogorov in a library. You may remember that in 1945 we also heard similar ideas expressed independently, and in very different mathematical form, by Heisenberg and von Weizsäcker who were then living under military restraint and who were brought to Cambridge at their request by an officer from the department of military intelligence. I remember that we spent some hours in my garden discussing their ideas and my impression is that they were new to me, although I did see the connexion with the theory of Kolmogorov. Later I heard that similar ideas had been put forward by Onsager. I certainly realized that if a statistically steady state could be established, with large eddies being supplied at a given rate and their energy finally disappearing owing to viscosity, a definite spectrum would result, but I had not thought how this could be expressed as a similarity proposition.

A lot of what was being said about the new theories must have been evident to G.I., and since the new aspects were not obviously connected with observable or practically relevant features of turbulence I understand now why he remained detached from these developments. Much of the subsequent published work on turbulence had a theoretical bias, and I doubt if he was ever tempted to take up active work on the dynamics of turbulence again. As in several other cases, he made the subject, and then left it before it became popular. G.I.'s glimpse of Kolmogorov's papers and his meeting with Heisenberg and von Weizsäcker showed him the direction in which postwar research on turbulence would go, but he was content to leave these further developments to others.

CHAPTER 13

Taylor's foreign peers in mechanics

I t is generally accepted that there were three giants in mechanics during the first half of this century; Theodore von Kármán (1881–1963), a Hungarian who worked at the University of Aachen for 18 years and then became Director of the Guggenheim Laboratory at Caltech in 1930;[1] Ludwig Prandtl (1875–1953), who was for many years Director of the Kaiser Wilhelm Institut für Strömungsforschung (later named a Max-Planck-Institut) at Göttingen;[2] and Taylor (1886–1975)[3]. Another prominent contributor to research in various aspects of mechanics in the thirties was Johannes Burgers (1895–1981), a younger man with a background in statistical mechanics, who was a professor at the Delft University of Technology from 1918 to 1955, when he migrated to the USA to take up a position at the University of Maryland.[4] These four men were near-contemporaries, and were all at their most active in the twenties and thirties. Collectively they dominated research on the mechanics of fluids and solids in that period. There is no space here for a description of their work, but it is of interest to see how G.I. interacted with these other leaders.

Evidence concerning G.I.'s relationships with von Kármán, Prandtl and Burgers is naturally sought mainly in the letters they exchanged. Unfortunately not a single letter from or to von Kármán

1. See figure 13.1(a), and *The Wind and Beyond: Theodore von Kármán*, Little, Brown & Co., 1967, for a readable autobiography of von Kármán.
2. See figure 13.1(b), and *Ludwig Prandtl: Ein Lebensbild, Erinnerungen, Dokumente*, by Johanna Vogel-Prandtl; Mitteilungen aus dem M.P.I. für Strömungsforschung Göttingen 1993, in which the author gives an account of certain aspects of the life of her father.
3. See figure 13.1(c).
4. See figure 13.1(d), and *Selected Papers of J.M. Burgers*, edited by F.T.M. Nieustadt & J.A. Steketee, Kluwer 1995, which is a combination of a biography and a collection of the more important papers by Burgers.

Figure 13.1 *Leaders in mechanics during the thirties.*

survives in Cambridge, although we know that G.I. and von Kármán
met at conferences, that G.I. visited von Kármán in Aachen in May
1930 in order to receive an honorary degree, that Leslie Howarth,
one of G.I.'s young Cambridge colleagues, spent the year 1937–8
working at Caltech with von Kármán on the dynamical equation for

the turbulent velocity correlation tensor, that G.I. and von Kármán were in contact over a question concerning the extension of vortex lines in turbulent flow during the late thirties, and that G.I. wrote two appreciative memoirs of von Kármán (1963b, 1973b). In the absence of letters I am reproducing here the brief memoir (1963b) that G.I. wrote on von Kármán's death in 1963 since it gives a good impression of the friendly relationship of these two very different great men, one an unassuming individualist and the other a colourful leader.

> For a man to remain an inspiring leader of scientific thought and endeavour for 50 years or more, as von Kármán did, is so remarkable that the editors of the *Journal of Fluid Mechanics* have asked me to give a brief appreciation of his work.
>
> Von Kármán was essentially an engineer whose thought extended over the whole range of engineering science. Though actively concerned with very large engineering projects he never ceased probing into the science which lay beyond them or speculating on their future effects on mankind. If the word broadminded were not so frequently misused in describing well-intentioned people with extensive, but often shallow, knowledge I would describe von Kármán as the most broadminded and most deepminded engineer of his generation. His quality appears even in his earliest paper published in Hungary when he was still at the Technical College in Budapest. This was really a student's exercise and described the falling of a round-ended rod standing on a table. Most of us would think that a rather unpromising subject but von Kármán extracted a surprising amount of interest from it, and made it a vehicle for some good mathematical analysis.
>
> After a few years with an engineering firm von Kármán went to Göttingen in 1906 where he came under the influence of Prandtl and started working on both the main subjects of his early scientific interests, fluid dynamics and the mechanics of solid continua. In his first work on elasticity he showed how the engineer's rules for calculating the strength of slender columns is related to Euler's theory of the collapse of an idealized strut. This paper is interesting because it reveals so clearly von Kármán's early engineering training in which empirical formulae played so prominent a part. Here he shows that the mathematician's idea is not so useless as some of his fellow engineers then believed, provided proper account is taken of the limitations which the

theoretician is forced to accept. In the course of his life von Kármán seemed to become more and more interested in ideal models from which practical detail had been omitted in order to expose the underlying physical principles involved. This seems to me to be one of the ways in which von Kármán has exerted so great an influence on engineering science.

It was during his career in Göttingen that he developed his completely idealized 'von Kármán Street' model of hydrodynamic resistance and showed how it is connected with what is observed in a water stream. In Göttingen also he collaborated with Max Born in calculating the vibrations of lattices and tracing their connnexion with specific heat. During this period his natural mathematical ability developed into a powerful tool which he used so successfully for the rest of his life.

In 1912 von Kármán went to Aachen where he started the Aeronautical Institute which rapidly became very famous and attracted students from all over the world. It was there I think that he developed his organizing ability. This did not take the form of making him a good organization man. I do not believe he was ever particularly skilled at the details of organization, but he seemed to have an instinct for seeing what problems were likely to be important from the engineering point of view, which had intrinsic intellectual interest and which of his students were likely to make a success of them.

After service with the Austro-Hungarian Aviation Corps during World War I he returned to Aachen, but the enmities raised during the war had a profound effect on him and he spent much time and effort in promoting international cooperation in science. One of the first fruits of these efforts was the starting of the International Congresses for Applied Mechanics. After a preliminary private meeting at Innsbruck in 1922 the first Congress was held in Delft in 1924, Biezeno and Burgers being secretaries.[5] These Congresses have had a most beneficial influence on applied mechanics as well as on the scientists attending them, particularly on scientists from countries whose

5. Burgers kindly invited G.I. and von Kármán to stay in his home during the Congress, thereby beginning a custom of friendly hospitality which continued for some years. Further Congresses have been organized, usually at four-year intervals, coming (by general agreement) under the aegis of the International Union of Theoretical and Applied Mechanics in post World War II years.

political systems tended to isolate them from their colleagues.

At the Congress meetings von Kármán always played an important role. As a Hungarian he was able to stand outside the national rivalries of French-, English- and German-speaking members which sometimes arose, and with his ready wit and ability to point out absurdity and inconsistency in an argument without hurting the feelings of the man who produced it, he was an ideal chairman. In this I think he was helped by his skilful use of what he described as the universal language, broken English, which gave an international flavour to everything he said.

During the period 1920-30 most of his effort was devoted to his Aachen Institute and his pupils, but some striking results of his own were published. To this period belong his turbulent boundary layer momentum equations and his logarithmic law for skin friction. The description of the laminar boundary layer on a rotating plate and of the mechanics of the rolling process in producing sheet metal were also published.

In 1930 von Kármán became Director of the Guggenheim Aeronautical Laboratory at Pasadena, California. Among the earliest products of his work with American students was his paper with N.B. Moore on the resistance of slender bodies at supersonic speeds. This was a pioneering study and was followed by papers from aerodynamicists all over the world. In 1937, following my own simple studies of the statistical theory of turbulence, von Kármán and Howarth showed how the theory could be much extended.

It was, I think, shortly after he went to California that von Kármán's great prestige began to be recognized among people who were not mechanical scientists. He was exceedingly quick at seeing and replying to points raised in discussion of non-scientific as well as scientific subjects, and he was able to express his thoughts on technical matters in an attractive way which non-technical people could understand once they had mastered the initial difficulty of interpreting his English. His advice began to be sought by government departments as well as by industrial concerns not only for its great scientific merit but also because they felt sure of its absolute integrity and because it was often expressed using apt illustrative analogies in a form which they could appreciate.

In California, I think less of his time must have been absorbed in administrative duties than it was in Aachen, for he obviously read very widely and the reviews which he published of the then existing state of knowledge were particularly valuable. They must have helped considerably in spreading his reputation.

Von Kármán was always an entertaining lecturer and sometimes he chose subjects of great general interest. One which he wrote with G. Gabrielli in 1950 had as its title 'What price speed?'. In it he compared all kinds of self-propelling objects from pedestrians to battleships using logarithmic diagrams representing horse-power per ton against speed. He found that there is a line with slope 2 to 1 above which all the curves lie and at their most efficient working the curves of the most efficient vehicles nearly touch this line.

Of von Kármán's work for government agencies and industry I cannot write with authority, but the fact that he was chairman of the U.S. Air Force Scientific Advisory Board indicates the trust which the Government of his adopted country placed in him. He never retired from active work, and his later years were devoted more and more to international activities. He founded the Advisory Group on Aeronautical Research and Development within NATO, and presided over its activities, spending most of the year at its headquarters in Paris. Though the word aeronautics appears in its title, its activities spread over a much wider field. It organized, for instance, scientific meetings on subjects which are developing rapidly; aerodynamic noise and . combustion are examples. One of von Kármán's last activities was to help in founding the International Academy of Astronautics, of which he became Director.

It would require more literary skill than I possess to convey to those who did not know von Kármán an impression of the delight and stimulation which his presence among us always called forth. We mourn the loss of an unselfish, wise and entertaining friend, and of a great citizen of the world.

Fortunately, we do have records of extensive correspondences conducted by G.I. with Burgers and Prandtl throughout the period 1924–37. I have received copies of the letters exchanged between Burgers and Taylor by courtesy of the Delft University of Technology and those between Prandtl and Taylor with the kind help of Professor E.-A. Müller, a successor to Prandtl as Director of the Institut für Strömungsforschung. Both sets of letters have gaps, although they are nearly complete for the period 1930–36. They provide an interesting although not always comprehensible commentary on the development of ideas about different aspects of fluid mechanics.

The letters from Burgers and Prandtl are formal and polite and type-written, those from Burgers being in English and from Prandtl in German, whereas those from Taylor are informal and friendly, and are written in English by hand. At one stage Prandtl begged G.I. to ask someone to transcribe his letters more clearly,[6] but there was no response from G.I. My general impression of the letters is that these men were open about their current research and willing to discuss each other's ideas without concern about priorities. A small selection of items of human or scientific interest from these letters is reproduced below. I take first the letters exchanged between Burgers and Taylor.

In the mid-twenties Burgers and Taylor had a common interest in measurements of velocity fluctuations in turbulent flow over a rigid surface, in a wind tunnel or a pipe in the case of Burgers and over the Earth's surface in the case of Taylor. The observations were made mainly by wind-vane anemometers and Pitot tubes, and Burgers and his colleagues were also beginning to use hot-wire anemometers. The measurements were crude but they facilitated the development of ideas about the essential mechanics of the different regions of a turbulent boundary layer; for example, the existence of a viscous-dominated layer between the wall and the fully turbulent layer was confirmed. Burgers showed G.I. a new argument giving the criterion for suppression of the turbulence in shearing flow of stratified fluid, and they discussed in a friendly and interested way several other topics, including a new formula for the virtual mass of an airship found by G.I. (see 1928d,e). As usual G.I. enjoyed any opportunities for adventure which came his way. He reports in a letter to Burgers dated 15 November 1929: 'This afternoon I have been for a 3-hour trip in the new airship R101. The manipulation of this enormous vessel in getting under-weigh and in landing is wonderful. Recently she stood a gale, when attached to the mast, in which the maximum speed in a gust was 83 miles/hour, and the crew necessary for handling the landing is only 14 men.' It seems likely that G.I.'s membership of one of the sub-committees of the Aeronautical Research

6. Everyone who worked with G.I. had to become accustomed to his handwriting, which was idiosyncratic although regular.

Committee at that time was the link between this trip and his calculations of virtual mass.

Burgers gave much thought to ways in which he could employ the techniques of statistical mechanics in an analysis of turbulent flow. For this it seemed to him to be necessary to find discrete entities of some kind of which the turbulent flow might be supposed to be composed. Burgers tried various possible entities for a two-dimensional incompressible flow, for example vortices, or the terms in a suitable expansion of the stream function, but was not satisfied with any of them. On describing one such attempt to G.I. he received a reply (dated 17 November 1929) which began: 'I have just been reading your interesting and valiant attempt to reduce turbulent flow to a statistical basis'. We need not read more of the reply to know that G.I. thinks the approach is flawed, because the word 'valiant' as used by the kind-hearted G.I. says just that.

A few years later G.I. introduced Burgers to his new ways of thinking about turbulent velocity fluctuations in terms of correlations between the values of the fluid velocity defined in the Eulerian manner at two different points. G.I. wanted to know whether Burgers had previously heard of anyone working out the kinematical consequences of an assumption of spherical symmetry, or isotropy, for the various correlations between spatial derivatives of components of the velocity vector (letter dated 2 May 1935). Burgers knew of no such work, and he looked forward to learning from G.I. about the dynamical role of these velocity correlations. Burgers seemed at this stage to be pessimistic of getting anything useful from the methods of statistical mechanics, which had little in common with G.I.'s new ideas on the statistical representation of velocity fluctuations. Burgers' last prewar letter written on 11 December 1935 said that he would try 'to get some insight into the relations which are of importance (viz. those determining the transfer of energy to smaller-scale motions) even if not for the actual hydrodynamical case, then still for some related example.' This intention led ultimately to formulation of the model equation

$$\frac{\partial u}{\partial t} + u\frac{\partial u}{\partial x} = \nu\frac{\partial^2 u}{\partial x^2},$$

now known as the Burgers equation, as a relatively simple nonlinear diffusion equation having solutions from which one might obtain insight into the fundamental processes governing real fluid turbulence. The possibility of a link with turbulence has remained uncertain, but the Burgers equation has taken on a fruitful life of its own in more recent years.

Burgers wrote one letter to Taylor during the war, on 3 May 1940, primarily to tell him that he had been elected as a foreign member of the Royal Netherlands Academy of Sciences. And there is a further phase of the correspondence between Burgers and Taylor between July 1945 and August 1948, mostly concerned with attempts by Burgers to arrange visits to Britain by Dutch scientists and students. Among Dutch scientists there was a yearning to learn what had happened scientifically during the war, and Burgers with typical unselfishness threw himself into the task, difficult at that time, of organizing contacts, and using his friendship with Taylor and Melvill Jones and others where possible.

The correspondence between Prandtl and Taylor began with a letter dated 25 April 1923 from Prandtl in which he raised for discussion the way in which small non-spherical particles make visible the motion of the fluid in which the particles are suspended. Prandtl had seen the note (1923d) in which Taylor described his observations of the tendency for ellipsoidal particles suspended in fluid in simple shearing motion to take up a preferred alignment, and he wondered whether this dynamical effect was relevant to the strange optical lustre of certain particles near the free surface of a moving liquid. Prandtl had experimented with a variety of particles for the purpose of demonstrating the generation of vortices in a simple open water channel. The first such demonstration, using finely divided iron oxide particles, accompanied his seminal lecture on boundary layers at the Mathematical Congress in Heidelberg in 1904, and later many illuminating photographs of the flow showing boundary-layer separation and generation of vortices were disseminated by published papers and slide projectors to students and teachers throughout the world. Prandtl sent Taylor a sample of a paste of iron oxide particles, but there is no letter of reply from Taylor in the collection.

On 16 May 1927 Prandtl gave the 15th Wilbur Wright Memorial Lecture in London on 'The generation of vortices in fluids of small viscosity'. The photographs mentioned above were projected, together with later ones showing the paths of particles of powder sprinkled on the free surface of the water. The first version of Prandtl's famous educational film on the generation of vortices was shown at the end of the lecture. Taylor invited Prandtl to visit Cambridge after the lecture in London, and in his later letter of thanks (4 June 1927) Prandtl brought their correspondence back to the topic of the semi-empirical transfer theories, which, as already noted, was central to turbulence research in the late twenties. In a gesture which perhaps owed more to friendship than to logic, Prandtl proposed that the rate of increase of mean momentum be written as $\alpha M' + \beta M''$, where M' and M'' are as defined in chapter 12 and α and β are adjustable constants, thereby catering for the possibility of each of the two transport theories being applicable in appropriate circumstances. However he concluded later that this formula had no physical justification.

In a significant letter dated 25 July 1932, following a gap of two years in the correspondence, Prandtl congratulated G.I. for having shown from observations, some of which were made at the National Physical Laboratory, that the fluctuating vorticity in the turbulent wake of a cylinder is primarily transverse to the mean flow whereas in turbulent flow over a rigid surface it is primarily parallel to the mean flow, a finding with clear implications for eddy transfer theories. This seemed to account nicely for the empirical applicability of the vorticity and momentum transfer 'theories' respectively . Prandtl went on as follows: 'It would be of great interest to me to hear whether you also are working in this direction (viz. comparing the measured properties of different turbulent flow systems with the predictions of the transfer theories). In this case it would be best to come to an understanding with you concerning further work about these matters. I feel so strongly the want of this understanding that I tried to write this letter in English to make it easier for you to read it.' G.I. to conduct his research according to some 'understanding'? Evidently Prandtl does not know that G.I. always went his own way. At all events, G.I. overlooks or ignores the request. Later in the same letter

Prandtl makes the generous admission that, 'In recent weeks I have studied your old papers (1915a) and (1921b) with the greatest interest; I think that if I had known about these papers, I would have found the way to turbulence earlier'.

In July 1934 Prandtl and von Kármán stayed with the Taylors at 'Farmfield' during the 4th International Congress of Applied Mechanics in Cambridge. G.I. was one of the organizing secretaries of this Congress (an unusual role for him), and he gave a general lecture on the strength of crystals of pure metals and of rock salt. They also met in Rome in 1935 at the Volta Conference on High Speed Flow (a landmark conference, to which G.I., Prandtl, von Kármán and Burgers all contributed).

In a letter to Prandtl dated 15 November 1935 which was mostly about combustion waves, G.I. broached a personal matter which he hoped would be regarded as confidential. G.I.'s former teacher, C.T.R. Wilson, had told him that candidates for the Nobel prize in physics were nominated by the past Nobel prize winners. This seemed unfair to G.I., because virtually all the past holders had won the prize for work on what he called 'atomic physics' and he thought it likely that this bias would be self-perpetuating. With unusual candour G.I. wrote, 'I feel very strongly that if the Nobel prize is open to non-atomic physicists it is definitely insulting to us that our chief – and I think that in England and USA at any rate that means you – should never have been rewarded in this way'. G.I. supposed that nothing could be done about the matter but he wanted Prandtl to know what he and 'many other people in many parts of the world are thinking about it'. The bias towards 'atomic physics' among winners of the Nobel prize for physics has continued, and no specialist in mechanics has yet been successful, although G.I. (and possibly others) was later nominated many times.

Further evidence of G.I.'s high regard for Prandtl may be seen in letters early in 1936 telling Prandtl that the University of Cambridge had agreed to offer him an honorary degree in June, and that he would again be welcome at 'Farmfield'. While in Cambridge for the conferring ceremony Prandtl gave a seminar on the lift on wings at supersonic velocities with many Schlieren photographs. He also visited the Royal Aircraft Establishment at Farnborough, and then flew

Figure 13.2 *Prandtl with (from the left) J.A. Haslam and B. Melvill Jones, from the Department of Engineering, and Taylor, in Trinity College in June 1936. Prandtl was in Cambridge to receive an Honorary ScD.*

home to Berlin from Croydon. The correspondence continued, less frequently, on a variety of topics relating to turbulence, up to May 1938 on Prandtl's side and January 1937 on G.I.'s side. Politics was beginning to interfere with international relations. In a memoir on von Kármán read on his receipt of the first von Kármán prize from SIAM, G.I. wrote (1973b) as follows:

> By the time the 4th Congress was held in Cambridge, England in 1934, the German Jewish members were having a bad time. Theodore was well out of it but was doing a lot for his unfortunate fellow countrymen. Prandtl, who was not Jewish, appeared to be completely taken in by the Nazi propaganda and when the news of Roehm's murder came while he was in Cambridge he refused to believe it and said the papers had invented the story. In 1938 however when the 5th Congress was organized in Cambridge, Massachusetts, by Theodore and Jerome Hunsaker, conditions had changed. Prandtl and my wife and I were staying with Jerome at his home in Boston. The

German delegation was strictly watched by political agents who had come as scientist members and Prandtl did not dare to be seen reading American papers. He used to ask my wife to tell him what was in them. After I had returned to Cambridge from the 5th Congress, I had a letter from Prandtl telling me what a benevolent man Hitler was and including a newspaper cutting showing the Fuhrer patting children's heads. I imagine the poor man did this under pressure from the propaganda machine, for other people told me they had similar letters from him.

Soon after the end of World War II Prandtl attempted to make contact with G.I. and to take up again the discussion of current research. In the recent biography of Prandtl already referred to in this chapter, the author (his daughter) reproduces excerpts in German from two letters to G.I. Here is an English translation of these excerpts:

28 June 1945

My Institute survived undamaged through the war. However, there is now a lot of damage because American soldiers were accommodated here for several weeks. Only after the beginning of June were we allowed to enter there. An Allied Committee told us what to do. We were allowed to make some repairs and to write reports for the Allied Committee. We were also permitted to continue certain projects which remained unfinished during the war and on which reports were expected. So far we have not been permitted to begin new work. Yet we hope soon to be able to investigate problems of a fundamental nature which were set aside during the war. We have enough such problems to last a decade!

10 Oct 1945

Any continuation of our research has hitherto been prohibited by the Director of Scientific Research in London. For a Research Institute whose mission is to extend the knowledge of its subject as far as possible this is a very hard demand. Indeed, we see many problems which await their solutions, for instance turbulence and near-sonic flows. Problems of meteorological and oceanographic flows in which density stratification and turbulent processes both play an essential part are also among such topics.

I do not know whether G.I. replied to these letters. It would surprise me if he did. He might have interpreted the letters as an appeal
from Prandtl for his support of the Institut für Strömungsforschung
in the chaotic conditions of Germany in 1945, and if so I doubt if
G.I. would have wished to intervene. If the letters were simply an
appeal for resumption of a friendly personal exchange, it was perhaps
too soon to expect that British people who had been engaged in a terrible war for nearly six years would be able to judge whether any of
their former German colleagues had been tolerant of Nazi-ism, unjust
though that may have been for some individuals. Prandtl died in
August 1953. I do not know whether he and Taylor met again after
the war. There appear not to have been any German delegates at the
Congress held in London in 1948, although there were some (not
including Prandtl) at the next Congress in Brussels in 1952.

CHAPTER 14

The universal defence consultant
during World War II

Taylor was probably at the peak of his scientific powers in the nineteen-thirties, and it was an immensely productive period for him. Several of the themes which had interested him throughout his life were carried to natural terminating points during this period, which was fortunate because the advent of war again put an end to research of his own choice for many years. Whereas during World War I G.I. had to offer his services to army and civil authorities who were uncertain about how scientists might be used, on this occasion his advice and help were greatly in demand from the armed services and many ministries and government agencies. World War II was, technically speaking, an exercise in applied classical physics (at any rate, until the appearance of the atomic bomb) on a vast scale, and the profound understanding of mechanics and physics that G.I. possessed fitted him perfectly for the role of consultant on the innumerable problems that arose. As one measure of the breadth of his advisory services we know that in 1940 he was a member of 15 committees and sub-committees of the Aeronautical Research Council, and many others under the aegis of the Ministry of Defence and the Ministry of Supply. There was no one in Britain with an equal skill in sizing up a novel scientific situation, uncovering the essential processes at work and formulating the real problem. I do not think the war years gave him much scope for exercise of his particular kind of creative originality, possibly because he was not brought into discussions of military and civil objectives at a sufficiently early stage, but he was much used, and was outstandingly effective, as an analyst and adviser on a multitude of problems.

For information about what G.I. did during the war, we have the memories of the very few living people who were in contact with him at the time and a large number of reports (included in the bibliography at the end of this book) written for various ministries or government committees and mostly published in the four volumes of his *Scientific Papers*. I recall the great difficulty that I had, as editor of these volumes, in finding out what reports he wrote for the many temporary war-time committees and in obtaining copies of them, and this makes me doubtful about the completeness of our record of his activities at this time. Perhaps I should explain here that, so far as I know, at no time in his life did G.I. employ a secretary or have his letters typed. The documentary evidence of what he did throughout his life consists mainly of letters and papers (including, of course, his own in published form) delivered to his house 'Farmfield', and since his filing system was rudimentary, and dependent more on his wife's wish to contain the papers in one room than on his need to find something later, I am sure there are some gaps. He did make an effort to retain one copy of every published paper by him in a set of boxes, but typed or duplicated reports, by him or by someone else, often remained in the envelopes in which they were delivered, and incoming letters were collected in large brown envelopes marked only with the year. Periodically Stephanie had a clearing-up operation which led to some documents being thrown out in order to make room for new ones, and relatively few of the letters and documents that came in before about 1960 have survived. I think it unlikely that any piece of work of real scientific value has been lost, because G.I. had a very good memory for his own investigations and took pains to put into publishable form any interesting ideas or developments. However, tracing the complete sequence of his activities during World War II, in particular those that helped other people's scientific work, is now virtually impossible.

We get a vivid glimpse of the conditions under which G.I. worked during the early years of the war from the following letter from G.I. to his friend Jerome Hunsaker at MIT. Hunsaker had the letter typed, no doubt for the benefit of members of his family unfamiliar with G.I.'s handwriting, and he sent a copy to G.I. in October 1963. G.I. is uncharacteristically emotional, even irrational, about the tactics of the German air force, possibly reflecting the general feelings of British people at that tense time shortly after the evacuation from Dunkirk.

13 Oct 1940, Farmfield

Dear Hunsaker,

I enclose a note written by a research student of mine, Mr D.C. Macphail, on some work done in my new wind tunnel just before the outbreak of war. Macphail is now working at Farnborough but he has had time to write up some of his work in the evenings after his work there. If you could publish this in the *Journal of the Aeronautical Sciences* I think it would be worthy of that excellent journal. The work really describes an extension of existing technique in turbulence measurement. The result is only what we might expect.

Stephanie and I often recall our stay with you in 1938 and wonder how you are. Whenever she gets hold of a new Wodehouse she is reminded of Mrs Hunsaker and hopes she is enjoying it too.

We are all pretty busy here. I have to travel about the country quite a lot and frequently go up to London to meetings and conferences. Things became a bit difficult in the early stages of the bomb dropping but it soon became clear that it wastes too much time to take cover whenever a warning is sounded. Now we find that we can get on with our business much better if we have a man on lookout on the roof who advises us when he sees a Nazi aeroplane so that we can get below. Casual bomb dropping at night and less casual bomb dropping by day is beastly enough but the thing that makes us really see red is when these things find a village or town where they are momentarily free from our air defence and come down and machine gun the people in the streets, women doing their marketing, children coming out of school. Of course they can only do that occasionally by gliding with engine off from a great height but whenever they get a chance they seem to take it.

Cambridge is still standing in spite of Dr Goebbels' announcement that it is a heap of smoking ruins. A good many bombs have been scattered around but we have only had two definite attacks. Fortunately none of the Colleges or old buildings were hit.

I am one of the few people who have not gone into one of the service establishments or aircraft factories. We still live in Farmfield though I am away a good deal. We have an air raid shelter but it is so unpleasant going out to it on cold nights that we have given up using it. We have however given up sleeping upstairs. We make our beds under our staircase because staircases often stand up when the rest of the house collapses.

I think things are not going too badly in the aeronautical field in this country and I expect you can appreciate that some of the work sponsored by our Aeronautical Research Committee during the pre-war period has turned up trumps.

Please give our warmest greetings to your wife and family.

Yours sincerely, G.I. Taylor

In 1938 the Home Office became concerned about the effects of aerial bombing in the event of war with Germany, and set up the Civil Defence Research Committee to advise on what bombs would do. G.I. was a member of the committee from the start, and quickly became drawn into investigations of the processes associated with the detonation of high explosives and the subsequent propagation of blast waves in air. Problems concerning the prediction of bomb damage, how to restrict it and how to increase it, grew enormously in importance as the war developed, and this became the main area of G.I.'s scientific work. The late Harry Jones, Professor of Mathematics at Imperial College, was closely associated with G.I. in much of this work and at my request in 1975 he compiled the notes that follow.

Explosions and blast waves in air (by Professor Harry Jones)[1]

Very soon after the outbreak of World War II Professor R.H. Fowler initiated certain thermodynamic calculations, relating to the explosion products of TNT, with the object of obtaining better quantitative estimates of the expansive effects of the product gases of the explosion. Not long afterwards, however, he left for an extended visit to Washington and arranged that the work should continue under the guidance of Professor G.I. Taylor. It was in these circumstances that, early in 1940, I found myself in a position to see much of Taylor's war-time work as it developed. In the early years of the war a good deal of Taylor's work was reported to the Civil Defence Research Committee of the Ministry of Home Security. He was also a member of the Physics of Explosives Committee (Physex) of the Ministry of Supply and he arranged that Dr W.G. (later Lord) Penney and I, who were colleagues at Imperial College, should join him on this committee of which later, following Sir Robert Robertson, he became chairman. He was an effective chairman but

1. Initially this was a contribution to my Royal Society Biographical Memoir of G.I. Taylor (*Biog. Mem.* **22**, 1976, pp. 563–633). It is reproduced here with minor changes only.

the meetings not infrequently resembled scientific seminars rather than formal business meetings; blackboards were often in use. Taylor's particular genius lay in creative activity rather than the formal procedures of administration. Most of the work contained in his reports was carried out in the sitting-room of his own house and problems and research projects would often be outlined on scraps of paper during train journeys between London and Cambridge.

In 1939 very little information was available on the mechanics of detonation in solids and on the resulting blast waves. There was of course a considerable body of general experimental data on the velocities of detonation waves in various types of high explosives, and the total energy released per gram of TNT for instance had been measured by exploding small charges in a strong and enclosed steel vessel. Information relating to blast waves included measurements of the time of arrival of the wave front at a specified distance and of blast-wave pressures estimated by a variety of gauges. On the theoretical side the mechanics of one-dimensional waves of finite amplitude and of shock waves in gases was well understood, and a partial theory of detonation in gases had been developed 30 to 40 years earlier. What called for urgent attention at the time was the application of this knowledge to gain useful information on the damaging effects of the blast waves which were to become an all too familiar aspect of life a few months later. It was a challenge which Taylor met with enthusiasm and great effect. His deep knowledge of the principles of fluid mechanics and a very sure feeling for problems which are quantitively solvable led him to produce a series of articles which did much to provide a quantitative background of knowledge in a field which had hitherto been the subject of judgements and more or less enlightened guesses. Many of these articles, though interesting by any standards, were only printed for the first time in the volumes of his collected papers, and there was other research which is not well documented. The following short account of some of this work, as it was done, may help to reveal the great contribution he made to the war effort.

The first two of Taylor's war-time reports (1939c, 1940d) were written with the object of clarifying the mechanics of the phenomena which the constructors of air-raid protection would have to contend with. He began by explaining the properties of shock waves in air, showing in practical terms the pressure, temperature and velocity relations, and went on to discuss the properties of the wave of finite amplitude which follows the shock front in a blast wave and which continuously changes its form. It was generally held that a blast wave was a detached disturbance consisting of a pressure wave followed by a wave of suction, but Taylor pointed out that this simple picture cannot be accurate because it fails to satisfy a basic relation between the velocity and the pressure which must hold in any wave which progresses in one direction only. He concluded that complicated reflexions must take place. In fact the calculation of the pressure distribution in a blast wave is beset with extreme difficulties because, as the wave weakens with distance travelled, the entropy production in the shock front decreases and therefore the air in the following wave has different values of the entropy at different points; in other words a single adiabatic pressure–density relation does not hold throughout the entire wave. Moreover since at that time the dynamical properties of the explosion products were quite unknown it was out of the question to attempt a direct calculation. What then could usefully be done? Taylor supposed that pressure–time measurements could be made at a distance of say 50 ft from a small charge, and then by workable approximations and assumptions, always in conformity with basic principles, he could determine the pressure–time relations at say 100 ft and 200 ft and even backwards at 35 ft. Such knowledge was useful because although damage near a bomb had to be accepted there was an extensive region in which protection was feasible and quantitative estimates of the impulse of a blast wave could assist design work.

Early in 1941 Taylor carried out an investigation into the possibility of detonation waves in three dimensions as distinct from the familiar one-dimensional detonation of gases in tubes. He examined the detonation of a spherical charge of TNT ignited at the centre, and analysed the motion of the product gases of the explosive behind the detonation front. In a well-known memoir written more than 30 years earlier E. Jouguet had come to the conclusion that such spherical detonation waves were not possible. It may be noted in passing that the nightly occurrences in Britain during the war years were no proof that Jouguet was wrong; most bombs are roughly cylindrical charges detonated at one end. Taylor came to the conclusion that spherical detonation waves are possible provided that infinitely rapid rates of change of velocity and pressure, exactly at the detonation front, are admissible. He pointed out that similar discontinuous changes are admissible in shock-wave theory when the structure of the front is not under consideration, and lead to results in agreement with observations. This work contained new dynamical theory and was later published (1950a). The basic idea underlying the treatment of the spherical case was that a solution of the dynamical equations could be found in which the velocity, the pressure and the density were functions of a single variable only, namely the ratio of the distance from the centre to the time measured from the moment of ignition.

Another of Taylor's fruitful ideas which originated in the period 1940–41 was the analysis of a detonating solid explosive confined by a heavy metal tube (1941c). The over-riding difficulty with calculations of a blast wave, whether in air or in the product gases of the explosion, is the non-uniformity of the production of entropy in a shock front. He therefore sought a practical arrangement in which this trouble did not arise. He found it in the configuration of a long cylindrical charge confined by a metal tube. In this case the gas flow behind the detonation front is determined essentially by Bernoulli's equation and the adiabatic relation between the pressure and the density of the product gases of the explosion. The rate of expansion of the tube, given the mass per unit length, can readily be calculated. It turns out that this arrangement is remarkably efficient as a machine for converting chemical energy into mass motion. G.I. mentions a figure of 96%, though in practice the fragmentation of the tube and other effects reduce the efficiency a little. This high efficiency is partly due to the nature of the adiabatic pressure–density relation of the products of combustion. Immediately behind the detonation front in TNT the density is approximately twice that of water and the internal energy consists to a large extent of potential energy arising from the repulsive forces between the molecules. The initial expansion therefore resembles that of a released spring rather than the expansion of a high temperature gas at more normal densities.

Following up his analysis of the cased cylindrical charge Taylor was able to solve, or at any rate throw a good deal of light on, what seemed a most intractable problem, namely the calculation of the velocity of fragments from an exploding bomb or shell (1943d). The casing of a long cylindrical charge expands into a conical shape of small semi-vertical angle just behind the detonation front and this angle and the velocity of the expanding case could readily be calculated and compared with flash photographs obtained by American workers. The interesting point here is that the case expands considerably without breaking up, as the flash photographs show, and during this time absorbs a large fraction of the energy released by the explosive charge. The interaction between the fragments and the explosion gases after fragmentation is less important than might have been guessed. Thus a good estimate could be obtained, not only of the fragment velocities, but also of their direction relative to the axis of the shell.

In 1942 Taylor adapted this configuration of the long cylindrical charge to determine realistic values of the pressures and velocities of shock waves from bare charges in different surrounding media, air and water in particular (1942f). Any attempt to do this directly by considering a spherical charge detonated at the centre is beset with great difficulties, some of which have already been mentioned and others are due to the superposition of direct and reflected waves. Taylor saw that one aspect at least of the problem could be solved by adapting the known solution for supersonic flow round a straight edge. Consider a detonation front propagated through a long cylindrical charge from right to left. An observer moving with the front sees the combustion gases streaming out from left to right and the surrounding air rushing past with the supersonic speed of the detonation waves. At the surface of the cylinder and close to the rim of the detonation front the conditions approximate to the flow round a straight edge. This flow both in the air and in the combustion products can readily be calculated in terms of a single variable, namely the angle which a radius vector centred on the edge makes with the axis of flow. These calculations gave the pressure in the shock wave close to the exploding cylinder and the direction in which the matter of the surrounding medium is projected forward relative to the axis of the charge. This is a good example of Taylor's power to select a problem susceptible of precise calculation which at first sight seems rather remote from the actual situation but throws much light on the essential mechanism of the process involved.

Taylor also made an important contribution to the theory of shaped charges with lined cavities. If a conical hole at the end of a cylindrical charge of explosive, with its apex on the axis, is lined with metal and then detonated from the other end it is found that a jet of liquid metal is shot forward along the axis with a speed that can exceed the velocity of the detonation wave itself. At the time Taylor applied himself to this problem in 1943 it had already been suggested by J. Tuck that this surprising phenomenon, so useful for armour piercing weapons, projected or hand-placed , in air or in water, could be explained on purely hydrodynamical principles. In a short report (1943c) just a few pages in length, Taylor elucidated the mechanics of the effect using nothing more recondite than simple geometry and Bernoulli's equation for the approximately steady flow relative to axes moving with the apex of the cone. It was Taylor's special gift to see how apparently complicated phenomena can be expressed in mathematical terms and to introduce a quantitative aspect into their description. Thus in the work described here where the magnitudes of the pressures, velocities and time intervals are far removed from everyday experience his thoughts still turned instinctively to simplified problems, amenable to mathematical description, and illustrative of the basic aspects of the phenomena in question.[2]

The war at sea and Operation Neptune

There was a committee on underwater explosions, called Undex, whose responsibilities were similar to those of Physex. Harry Jones and Bill Penney were members of Undex, and G.I. was also the maestro of the work of this committee. They quickly found that knowledge of blast waves propagating under water was just as meagre as that of blast waves in air. Bill Penney wrote as follows about the early work of Undex:[3]

2. A more extensive discussion of hollow shaped charges may be found in the later paper (1948e). The resulting jet is sometimes referred to as the Munroe effect.

3. Notes by the late Lord Penney in the remainder of this chapter, like those of Harry Jones and Arnold Hall, were originally contributions to my Royal Society Biographical Memoir of G.I. Taylor.

When I showed G.I. the results of my first calculations on the underwater blast wave, I asked if there were any experimental data. He looked a little sheepish and said he would try to get some for me. A few days later he produced a dirty piece of tracing paper. I had seen better drawings but what it showed was the pressure–time record fairly close to an underwater explosion. A few days later, I was allowed to see the source. I was very surprised to find that it dated back to the end of World War I, and was in a report by Hilliar. This was the best information available at that time.

Hilliar was also very much interested in the surface plume and he relates that the first phenomenon that can be seen when a charge explodes at some distance underwater is the sudden appearance, followed quickly by the disappearance, of a dark circle on the surface. I asked G.I. if he knew what caused the circle. He said he had thought about it and although he did not know the explanation he guessed it was due to an instability – drops might be thrown off the surface (or, more precisely, left behind by the surface returning to its equilibrium position). I asked if the circle could be due to cavitation bubbles in the reflected wave of tension and he said that this was the other possible explanation. I still do not know which is correct, but the point of the story is that a few days later G.I. told me that he had solved the problem of the behaviour of a disturbance to a plane interface between two fluids of different density in a field of acceleration. It is strange that his simple calculation of the relation between the exponential rate of amplification and the wavelength of the disturbance has brought his name to many fields of science, and more people know about Taylor instability than know about his marvellous work on much more profound problems.

An important practical problem referred to Undex and studied by G.I. was the effect of an underwater explosion on structures (1941d, 1942h). Knowledge of the damage done by such an explosion was needed for the determination of the effectiveness of weapons such as depth charges. Harry Jones wrote as follows:

The results are of considerable scientific interest and may be rather surprising to those not familiar with this field. For example, although the pressures measured by piezo-electric gauges around a submerged explosion showed no great difference whether placed vertically above the charge or at an equal horizontal distance, the damage suffered by steel plates placed horizontally above the explosion was much more severe than that suffered by the same plate placed vertically at the same distance and at the same depth as the charge. Several writers had, of course, studied the motion of water in the neighbourhood of an explosion, but Taylor's analysis especially clarified the mechanics of the effects just described. The greater damage caused in positions above the explosion was shown to be due to the combined action of two effects arising from gravity. Firstly, the pressure of the water causes the gas bubble to contract and then to expand a second time owing to the compression of the gas. Secondly the bubble rises carrying with it a considerable momentum in the water which becomes concentrated in a small volume during the contraction. It is the combination of these two processes which can produce such large effects.

The mathematical analysis of such a situation necessarily involves approximations and some assumptions, and Taylor always sought to complement analysis by experimental investigation. On this occasion the experiments were made with the assistance of R.M. Davies (1943a). Together they carried out small-scale laboratory experiments in which the explosive was simulated by an electric spark. Taylor had previously shown that the pressure of the atmosphere at the surface of the water plays a significant role, and for small-scale experiments it was necessary to reduce the surface air pressure by a considerable amount. The practical value of such work lies in the verification of various scaling laws based on arguments of dynamical similarity.

Another area of work on which G.I.'s experience of underwater blast waves was brought to bear (so I learn from Arnold Hall) arose from the sweeping of magnetic mines. When the Germans introduced the magnetic mine, it took the Allies by surprise, and a means of sweeping which could be produced rapidly was needed urgently. As a result of this some aircraft were prepared with a magnetic coil mounted on the wings which proved capable of dealing with the problem, and which were brought into service very quickly. Of course, the sweep was essentially a narrow one, and the ship-borne version which was subsequently introduced could cover a much bigger area – however, the ships could not be provided quickly, whereas the aircraft sweep could. One of the problems with the aircraft sweep was to ensure that the aircraft was not itself blown up by the mine it detonated. G.I. was consulted about the nature of the blast waves from an underwater explosion passing through the water and subsequently through the air, and the calculation of the safety distances needed, although there are no surviving notes on the work.

G.I. was naturally involved in many different aspects of Operation Neptune, the landing of Allied forces on the coast of Normandy in June 1944. Bill Penney told me the following story about some of the forward planning:

> A naval officer came to see me in late July 1943, and asked if I could meet him in Professor C.M. White's hydraulics laboratory at Imperial College the next morning. The officer said he had a very secret assignment about temporary harbours for Operation Neptune. The next day, I went to White's laboratory and there was Taylor with the naval officer and White. A metal tube was across the bottom of a wave tank, and compressed air was being fed into the tube and was escaping through a series of fine holes in the tube. This, said the naval officer, is a model of a bubble harbour. Certainly, the curtain made by the rising bubbles reduced the incoming waves almost to nothing. White had sound instincts and said that he thought the scheme was impractical because it needed so much power. Taylor, in the next few days, set out calculations which confirmed this conclusion and which really did carry conviction. (Taylor was able to analyse the buoyant-plume part of the bubble-harbour problem quickly because he had earlier done similar calculations on temporary clearance of fog from airfields by lines of oil burners in 1942. The basic model is a turbulent plume above a line source of buoyancy, with the mean inflow velocity at any height being assumed to be proportional to the local mean vertical velocity in the plume.[4]) This was important because the Quebec Conference was to take place only three weeks later. The Navy was determined to say how they were going to provide artificial harbours to facilitate the invasion of Europe the following year, and the Army was equally determined to make the harbours their own way. Taylor had killed the idea of the bubble harbour just in time but the Navy then substituted floating devices. Churchill and

4. See the later paper (1956d).

Roosevelt accepted the Army's caissons and blockships for the inner harbour and the Navy's floating devices for the outer harbour. I was at the conference and watched with incredulity because all that was known about floating harbours was based on a few experiments in a wave tank with air-filled rubber bags. I remember helping to calculate that we would need so many millions of square yards of canvas, several thousand tons of rubber and that Dunlop's could just get the job done in time. The United States top brass agreed to supply, including several liberty ships to carry the stuff to Britain, and President Roosevelt personally approved the commitment to transportable prefabricated harbours.

The success of Operation Neptune was critically dependent on favourable wind, wave, moon, tide and cloud conditions, and it was important that meteorological advisors to the military leaders should be able to make an approximate forecast of the weather for six or seven days ahead. It was of course impossible in 1944 to make such medium-range forecasts reliably; and even today, with enormous computing resources, the margin of error is wide. The Air Ministry asked the high-level Meteorological Research Committee to advise on the best available methods, and this committee set up a sub-committee under G.I.'s chairmanship to examine 'further work on various methods of extending the time span for which forecasting for several days ahead was possible'. However, the military commanders for Operation Neptune appear simply to have asked the best meteorologists to use their experience, insight and wisdom, with tried and tested methods.

In the event weather variations played quite a dramatic role in the invasion of Normandy, as told in a record recently compiled in commemoration of the 50th anniversary of D-Day.[5] During the period of a few days beginning Monday 5 June 1944, the tide and moonlight were favourable for a landing, and this date was General Eisenhower's provisional advance choice for D-Day 'subject to last-minute revision in the event of unfavourable weather'. But on the evening of Saturday, 3 June, in fine weather, the advice from the meteorologists was that the weather 30–48 hours later on the

5. *With Wind and Sword: the Story of Meteorology and D-Day 6 June 1944*, by S. Cornford, Meteorological Office, 1994.

other side of the English Channel would be 'below acceptable levels' for landing craft and aircraft. The assault was therefore postponed for 24 hours, and convoys already at sea were ordered to put about. On Sunday 4 June, the situation was assessed again. There was driving rain at Eisenhower's headquarters, but the meteorologists believed that a ridge of high pressure developing behind the advancing front could provide a temporary window in the unsettled weather over the assault area just for the critical hours of Tuesday 6 June. Eisenhower was advised that he could count on fine weather during the morning and that it might last through the afternoon. On the basis of that limited forecast the decision to invade Normandy on Tuesday 6 June was taken. The sea was still rough and it was an uncomfortable 17 hours crossing for men in landing craft, but this worked to the advantage of the Allies in that the German commanders concluded the conditions were unsuitable for invasion. On the next occasion when the tide was favourable, two weeks later, the Channel weather was the worst for 20 years.

Development of the atomic bomb

If one physicist said to another, during the early forties, that he expected to be away on a trip soon, and then said in reply to an enquiry about where he was going, 'I can't tell you', it was a reasonable inference, understood among physicists, that he was off to Los Alamos to work on the highly secret Manhatten Project, code name for the development of the atomic bomb. A great concentration of scientific talent was assembled in the scenic town in the mountains of New Mexico, and it was inevitable that G.I. should be included. G.I. paid three visits of several weeks duration to Los Alamos, the last one being in the summer of 1945 to witness the first test of the atomic bomb on 16 July. G.I. acted as a general scientific consultant, and according to his colleague Bill Penney his advice was sought on such matters as the passage of a detonation wave from one explosive to another; the possibility of a jet when a detonation front moves across two pieces of explosive not fitting closely together; the mechanics of a converging detonation wave; and the theory of a detonation wave in an explosive impinging normally or at an angle to a plane interface with a metal. He (and others) pointed out that in some of the proposed configurations for the implosion phase of a plutonium-bomb explosion there would be a 'Taylor instability' which would wrinkle the spherical wave front and start the fission process too soon.[6] And as everyone knows, he obtained the famous similarity solution for the spherical blast wave from a point source of energy. This last contribution warrants a fuller description.

The story of G.I.'s involvement in the development of the atomic bomb began in April 1941 when he was consulted by G.P. Thomson (son of J.J.).[7] Thomson was chairman of the recently established M.A.U.D. committee (a camouflage name) and he told G.I. in strict confidence that it might be possible to produce a bomb which would release a very large amount of energy by nuclear fission. The question on which G.I.'s considered opinion was sought concerned the mechanical effects of such an explosion. The common chemical explosive bomb generates suddenly a large amount of gas at a high temperature in a confined space. The new bomb was expected to release a very large amount of energy in a small unconfined space without the generation of gas: would the mechanical effects be similar in the two cases?

In June 1941 G.I. produced a report for the Civil Defence Research Committee (1941e) in which he analysed the propagation of a spherical blast wave emanating from the explosion. He was able to obtain remarkably simple results for the trajectory of the blast wave, essentially by the use of two procedures, both time-honoured in fluid mechanics. One is to idealize a problem so that processes or factors expected to be of little relevance are ignored; in the present case G.I. assumed that a finite amount of energy E is released at the initial instant at a point, the blast wave is spherically symmetrical, the air pressure outside the blast wave is negligible (by comparison with that immediately inside it) and the air moves adiabatically, all quite reasonable. The other procedure is to identify all the relevant physical or geometrical parameters occurring in the data of the problem, here

6. It is sometimes said, incorrectly, that this was the context in which Taylor instability was first recognized and analysed; see the note by Bill Penney reproduced earlier in this chapter.

7. The source of most of my information about the role of British scientists in the development of the atomic bomb is the very readable and authoritative account by the historian Margaret Gowing, *Britain and Atomic Energy 1939–1945*, Macmillan & Co., 1964.

the atmospheric air density ρ_0, the constant total energy E and the adiabatic index γ, and to exploit the need for dimensional correctness of the expressions for the dependent variables, which here are the velocity and pressure and density of the air as functions of the radial coordinate r and time t measured from the instant of release of the energy.

Now this formal frame-work of the problem shows that length and time scales occur in the data of the problem only in the combination E/ρ_0. This is important, because it follows that a length characteristic of the flow field, such as the radius of the blast wave front (R say), is a function of t of the simple form

$$R = C(t^2 E/\rho_0)^{\frac{1}{5}},$$

where C is a dimensionless constant determined by γ. As in so many other of his investigations, G.I. was led in this way to postulate the existence of a *similarity* solution of a suitably idealized form of the problem, in which the spatial distributions of the flow variables are of the same form at different times, differing only in the scales. The various partial differential equations in r and t governing the flow can now be transformed into ordinary differential equations in the similarity variable $r^5 \rho_0/t^2 E$. Further analytical progress is difficult, especially as the entropy of the gas (and hence γ) varies across the shock front by an amount which depends on the shock strength, but the difficulty is of a lesser order in consequence of the possibility of transforming to similarity variables.

The detailed analysis enabled G.I. to estimate, in answer to the M.A.U.D. committee's initial question, that about half the energy released is degraded into heat, showing that an atomic bomb would be about half as efficient, as a producer of blast, as conventional chemical explosive – contrary to the more pessimistic estimates being made by some specialists in the USA, G.B. Kistiakovsky in particular.

It later became known that in the USA, J. von Neumann had also found the similarity solution,[8] and by coincidence submitted his work to a government committee also in June 1941; and for good measure the similarity solution was also published after the war by

8. The point source solution. *Collected Works, vol. VI*, pp. 219–237, Pergamon, 1963.

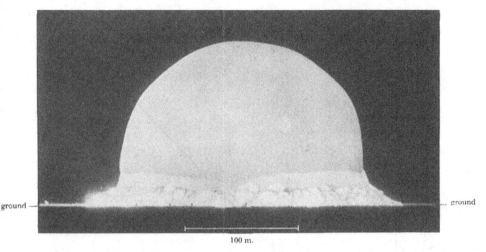

ground ——   ground

100 m.

Figure 14.1 *The ball of fire at t = 15 msec, showing the sharpness of its edge. (From 1950d)*

three other authors.[9] There was however a characteristic difference in G.I.'s work – he was not content with a purely theoretical investigation of the problem.

In 1947 photographs of the first atomic explosion in New Mexico on 16 July 1945 were released by the US Army and published in newspapers and magazines all over the world, and aroused great interest (see figure 14.1). A sequence of photographs of the 'fire ball' (the surface of which can be assumed to coincide with the blast wave front at times not too large) taken at different times after the explosion was now available, and G.I. realized that this sequence would enable him to test the validity of the similarity solution. Figure 14.2 shows his log-log plot of the observed radius of the fire ball as a function of time, and, for comparison, a straight line with the slope expected from the similarity solution. Furthermore, since the constant value of $Rt^{-2/5}$ is known to be equal to $C(E/\rho_0)^{1/5}$ and C is a known func-

9. L.I. Sedov, Propagation of strong shock waves, *Appl. Math. Mech.*, **10**, 1946, pp. 241–50; R. Latter, Similarity solution for a spherical shock wave, *J. Appl. Phys.*, **26**, 1955, pp. 954–960; and J. Lockwood Taylor, An exact solution of the spherical blast wave problem, *Phil. Mag.*, **46**, 1955, pp. 317–20.

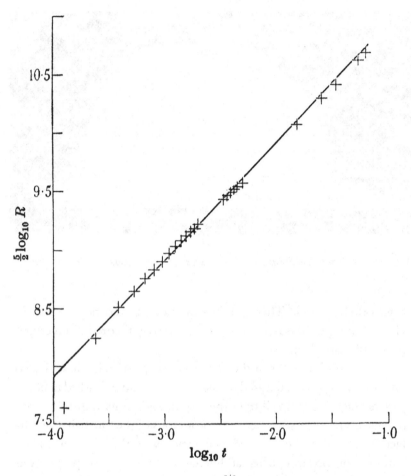

Figure 14.2 *Logarithmic plot showing that* $R^{5/2}$ *is proportional to t. (From 1950d)*

tion of γ, the value of which can be estimated, he thus had a way of estimating E, the energy of the explosion (more precisely, the part of the energy not radiated outside the blast wave front). The value of γ in reality is likely to vary with position in the fire ball, partly because an air molecule experiences dissociation and changes of vibrational energy at high temperatures and partly owing to radiative transfer of energy, but G.I. reasoned that since the observations show $Rt^{-2/5}$ to be quite closely constant these different sources of variation of γ evidently cancel each other, 'leaving the whole system to behave as though γ has an effective value identical with that which it has when none of them is important, namely $\gamma = 1.40$'. I have quoted G.I.'s own words (1950d) in order to show the intuitive basis of his selection of the value 1.40 for γ. The choice is important because the variation of E with the assumed (constant) value of γ is significant, as the following table shows.

γ =	1.20	1.30	1.40	1.667
E (in equivalent kilotons of TNT)	34.0	22.9	16.8	9.5

G.I.'s belief that the effective value of γ was 1.40 thus gives E as roughly 16.8 kilotons, in fair agreement with the figure of 20 kilotons announced by President Truman and based on measurements of pressure and air velocity and temperature on fixed instruments behind the blast wave generated by the test bomb.

Soon after G.I. saw the photographs of the first atomic bomb explosion he sought, and received, permission to publish in the Proceedings of the Royal Society his report of June 1941 to the Civil Defence Research Committee (1941e), which he called Part I, and a new paper, called Part II, in which he compared his theoretical results with the published photos and estimated the energy released by the bomb (1950c, d). This estimate of the yield of the first atom bomb explosion caused quite a stir, since only an estimate based on classified measurements made near the bomb had yet been announced, and G.I. was mildly admonished by the US Army for publishing his deductions from their (unclassified) photographs.

G.I. on a more public stage

The successful test of the plutonium bomb with an implosion firing mechanism in New Mexico on 16 July 1945 prepared the way for military use of the bomb, and a uranium bomb with a gun-type firing mechanism was dropped on Hiroshima on 6 August and another on Nagasaki on 9 August. On 6 August 1945 the secret was out, and the world learned of the existence of an awe-inspiring new weapon. Everyone wished to know more about this terrible weapon which killed 64 000 people in Hiroshima and which clearly would affect the fate of nations in the future. There was also a moral issue; was it necessary to kill so many people in order to demonstrate what the bomb could do? And if Hiroshima was a crime against humanity, Nagasaki was doubly so. In anticipation of this intense interest the US Army gave permission to observers of the first test of the bomb in New Mexico to describe their experience. The British Broadcasting Service arranged a group of radio talks on 'The genesis of the atomic bomb' and G.I. contributed one on 'Trying out the bomb' which was published in the *The Listener* (1945e). Here is the text of G.I.'s talk:

> I was one of the group of British scientific men who worked at Los Alamos in New Mexico, where most of the recent experimental work on atomic bombs was carried out, and I saw the first bomb explode. Before I tell you about this I ought to say that I have witnessed many ordinary bomb trials. In such trials the kind of result to be expected is always known beforehand and the trial is designed to find out just how much damage the bomb will do. The first atomic bomb test had to be approached with a totally different outlook because it was not possible to make any previous experiment on a smaller scale. None of us knew whether we were going to witness an epoch-making experiment or a complete failure. The physicists had predicted that a self-propagating reaction involving neutrons was possible and that this would lead to an explosion. The mathematicians had calculated what mechanical results were to be expected. Engineers and physicists had set up apparatus rather like that used in testing ordinary bombs, to measure the efficiency of the explosion. But no one knew whether this apparatus would be needed, simply because nobody knew whether the bomb would go off.
>
> Our uncertainty was reflected in the bets which were made at Los Alamos on the amount of energy to be released. These ranged from zero to the equivalent of eighty thousand tons of T.N.T. Those of us who were to witness the test assembled during a late afternoon in July at Los Alamos for the two-hundred-and-thirty mile drive to the uninhabited and desolate region where the test was to be made. We arrived about three o'clock in the morning at a spot twenty miles – notice it was twenty – from the hundred-foot tower on which the bomb was mounted. Here we were met by a car containing a radio receiver. Round this we assembled, listening for the signal from the firing point which would tell us when to expect the explosion. We were provided with a strip of very dark glass to protect our eyes. This glass is so dark that at midday it makes the sun look like a little

undeveloped dull green potato. Through this glass I was unable to see the light which was set on the tower to show us where to look. Remember, it was still dark. I therefore fixed my eyes on this light ten seconds before the explosion was due to occur. Then I raised the dark glass to my eyes two seconds before, keeping them fixed on the spot where I had last seen the light. At exactly the expected moment, I saw, through the dark glass, a brilliant ball of fire which was far brighter than the sun. In a second or two it died down to a brightness which seemed to be about that of the sun, so, realising that it must be lighting up the countryside, I looked behind me and saw the scrub-covered hills, twenty-two miles from the bomb, lighted up as though by a midday sun. Then I turned round and looked directly at the ball of fire. I saw it expand slowly, and begin to rise, growing fainter as it rose. Later, it developed into a huge mushroom-shaped cloud and soon reached a height of forty thousand feet.

Though the sequence of events was so exactly what we had calculated beforehand in our more optimistic moments, the whole effect was so staggering that I found it difficult to believe my eyes, and judging by the strong ejaculations from my fellow watchers other people felt the same reaction. So far we had heard no noise. Sound takes over one and a half minutes to travel twenty miles, so we next had to prepare to receive the blast wave. We had been advised to lie on the ground to receive the shock of the wave, but few people did so, perhaps owing to the fact that it was still dark and rattle-snakes and tarantulas were fairly common in the district. When it came it was not very loud, and sounded like the crack of a shell passing overhead rather than a distant high-explosive bomb. Rumbling followed and continued for some time. On returning to Los Alamos I found that one of my friends there had been lying awake in bed and had seen the light of the explosion reflected on the ceiling of his bedroom, though the source of it was over a hundred and sixty miles away in a straight line.

It has been my good fortune in three prolonged visits to Los Alamos to see the important contributions which the mostly young and so far unnamed British scientists have been making to the success of the project. Measurements of the blast are always made when new types of high-explosive bomb are tested and much the same kind of apparatus was used in the atomic bomb trial. It was from measurements of this kind rather than from nuclear measurements that the figure which has been given out as the equivalent amount of T.N.T. was deduced. Several British scientists were included in the group whose business it was to make these assessments. In conclusion I should like to say that the American, British and Continental scientists working in Los Alamos were fully alive to the effect which their work might have on human destiny and frequently discussed among themselves the ways in which the bomb might be used for the benefit as well as to the detriment of mankind. One frequently heard expressed, for instance, the hope that the immense destructiveness of the new weapon would so strengthen some really international authority that no nation would care to resist it.

Some of the scientists engaged in the work at Los Alamos were touched very deeply by the moral questions associated with the development and use of the bomb, but I do not think that G.I. was among them. He said later that like most scientists he was horrified by the dropping of the bomb on a city in Japan and even more when a second bomb was dropped, but he regarded that as the responsibility of the politicians. His BBC radio talk about the test of the bomb in New Mexico, given only days after the bomb was dropped on Hiroshima, was about the novel technical aspects and the drama of the first flash, the growing ball of fire and the mushroom cloud at dawn in the desert, and I think a number of people felt it was inappropriate to make an entertainment out of a matter about which at that moment they could only feel anguish. G.I.'s response to the wonderful physics of the bomb was innocent and natural, like his enjoyment of aeroplanes, and I doubt if he would have shared their concern. He was not reflective, and moral or philosophical issues did not often engage his mind.

There was one further episode in G.I.'s involvement in the development of the atomic bomb, this one being involuntary and rather bizarre. In 1970 a first novel by a young post-doctoral fellow in bioengineering at Harvard, Thomas McMahon, was published by Macmillan under the title *A Random State* in the UK and *Principles of American Nuclear Chemistry: a Novel* in the USA. The novel is concerned with the personal lives of some scientists working on the bomb at Los Alamos, as narrated in later life by the teen-aged son of one of the scientists named McLaurin. It is a work of fiction, but some of the scientists in the novel have names approximating to those of well-known physicists and McLaurin is identifiable as G.I.

by the author's description of his work on the blast wave from the bomb. This work on the blast wave is the only connection between the fictional McLaurin and the real G.I., but G.I. was incensed when he heard about the existence of the book and read a short review in *The Times*. G.I. objected especially to the fact that McLaurin had left his wife and was living with another woman at Los Alamos, and expressed his indignation in this typed letter to the publisher.

The Chairman, Macmillan Publishing Company,
10-15 St Martins Street, London W.C. 2

Farmfield, 1971

Sir,
   I have recently had letters from American friends telling me of a novel entitled *Principles of American Nuclear Chemistry* by Thomas McMahon in which I and a number of other scientific men appear as unpleasant people, mostly under thinly disguised names such as Orr – Nils Bohr, Ferrini – Enrico Fermi, Selina Meisner – Lise Meitner, etc. Your firm has published the same book with the same pagination under the name *A Random State*. Most of those mentioned unfavourably are dead and so cannot be libelled, but I am still alive. The identification of me with the father of the Protagonist in the book is unmistakable as you will see if you compare the description of two of my papers,[10] particularly the summaries, which I enclose herewith, with pages 80–82 in your book. My specific objection is that I am described as 'living in sin' at Los Alamos, a statement for which there is no foundation though it might have embarrassed my wife if she had still been alive. *The Times* reviewed the book on January 14th and you can appreciate the impressions that the book made on the reviewer who describes us as 'unpleasant, foul mouthed, sex obsessed stupid men in all areas outside their work'. This description of people like Nils Bohr, Fermi, Oppenheimer and Ernest Lawrence is so crude and untrue that no one who had the privilege of knowing them would be deceived by it, but it gives a completely distorted and untrue picture of us to the general public.
   I fear that nothing but the withdrawal of the book would be of any use in rectifying the wrong that your book has done to us, but I feel that you owe us some kind of apology. Your reader should never have asked you to print the book.

Yours faithfully, Sir Geoffrey Taylor, F.R.S., O.M.

10. The pair on the blast wave from a very intense explosion.

G.I. was 85 at this time, and was understandably upset by the dero-gatory and crude terms in which the (unknown) *Times* reviewer says the author describes the scientists. I have read the book, and in my view the scientists are not described in the book in the manner alleged by the *Times* reviewer. I believe the reviewer greatly exaggerates the unpleasant aspects of the scientists presented by the author. The *Times* reviewer should also have explained that the literary purpose of the portrayal of some of the scientists as coarse was to make a con-trast with the purity and beauty of their scientific research. The *Times* reviewer's opinion of the novel was in fact generally favourable. He described the book as being 'of extraordinary power and interest' and said it paints 'a grim picture, finely conceived.' C.P. Snow (who begins his review with the sentence: 'This is the best first novel I have read for years'), J.B. Priestley, and George Steiner and other well-known critics praised the book in long and serious reviews. If it had not been for the intemperate and inappropriate remarks of the *Times* reviewer, G.I. might have paid no attention to the book. Both the author and the publisher replied apologetically to G.I., saying that none of the characters in the novel were intended to represent real people and expressing surprise that G.I. judged he may be in any way represented in it.

CHAPTER 15

The aftermath of Taylor's war work

For several years after 1945 G.I.'s scientific work continued to be connected with practical problems which had arisen during the war. Many interesting phenomena had been revealed by the extremely wide range of conditions associated with military operations, and the imperative need for quick answers to novel problems had given a tremendous impetus to mechanical science. But the war years allowed too little time for investigation in depth or for following up interesting side-lines, and by 1945 G.I. had accumulated a pile of partially understood problems and ideas for further work. For a time therefore he devoted himself to clearing up the unresolved questions and to preparing for publication the many important but incomplete pieces for work which had initially been described in confidential war-time reports. He continued to hold the Yarrow Research Professorship awarded by the Royal Society, and so was free from teaching and administration to pursue his research interests.

Now that I understand his pragmatic tendency ('opportunist', as Southwell once aptly described it) to follow up any interesting questions which came to his notice regardless of their origin, it does not seem surprising to me that after 1945 he should continue to work on problems thrown up by the war. Certainly there were opportunities for fruitful research, as the following examples of G.I.'s investigations will show.

The war-time concern with the effects of explosions stimulated research on stress waves in solid materials, although it appears that G.I. had been interested in this topic even before the war. In 1946 he gave the James Forrest Lecture to the Institution of Civil Engineers on 'The testing of materials at high rates of loading' (1946a), and included in it the results of a number of investigations carried out during the previous nine years. The interpretation of tests of materials subjected to high rates of straining is made difficult by uncertainty concerning the effects of inertia of various parts of the machine holding the specimen and of the specimen itself. It is a field which puts a premium on ingenuity in the design of apparatus, and was an ideal choice for G.I. In the prewar work, carried out with H. Quinney, measurement of both stress and strain in a specimen beyond the yield point was achieved by attaching the specimen to a ballistic pendulum and firing a succession of lead bullets into a light anvil which pulled on the other end of the specimen. G.I. made a preliminary study of even higher rates of loading by projecting a cylindrical specimen at a hard steel plate, and worked out a formula which allows the dynamic yield stress to be deduced from measurement of the squashed cylinder after impact. This became of interest during the war in connection with the design of projectiles, and extensive tests were conducted at the Road Research Laboratory and described by Whiffin in a paper published later together with one on the theory by G.I. (1948b).

Perhaps the most important of the contributions by G.I. reported in his Forrest Lecture was his analysis of stress waves in a plastic material. In 1940 he saw a report of some observations on what happens when a bullet hits a wire, and realized immediately that the observed critical velocity of impact, beyond which the bullet breaks the wire, is associated with a critical velocity of longitudinal waves of plastic deformation (controlled by resistance to shear rather than by compressibility) in the wire and not, as the writer of the report had supposed, with the static strength of the wire. G.I. devised a theory of such plastic waves, and applied it to the propagation of waves in the Earth from an explosion close enough to the surface for the vertical displacement of the Earth to correspond to the lateral displacement of a bar subjected to end loading (1940a). The problem also turns up in a consideration of the stress produced in arresting wires stretched over the deck of an aircraft carrier during deck landings. The effect of the transverse blow is to lengthen the wire and so give rise to a longitudinal wave in the wire. This leaves the wire in a state of constant stress which permits a more slowly moving V-shaped lateral wave, like that in a stretched string, to travel outwards from the point of impact at uniform velocity. Two years later von Kármán independently tackled the same problem of propagation of a plastic wave in a stretched wire suddenly loaded at one end, with a different formulation in terms of Lagrangian variables, and obtained similar results (see 1942a).

G.I.'s involvement in the war-time committee on underwater explosions led to two important postwar publications. One was a study of the dynamics of large air bubbles rising through liquid (1950b). During the war G.I. had collaborated with R.M. Davies on a number of investigations of undersea explosions and their effects (and on some other problems – Davies was a skillful experimenter and they worked well together), and they had noted in particular that the initially spherical hollow produced by an underwater explosion subsequently becomes flattened on the underside (1943a). These war-time investigations were concerned mainly with the initial oscillatory phase of the motion of the cavity, and after the war Taylor and Davies explored the later phase of steady rise of the gas bubble. By taking spark photographs of bubbles large enough for surface tension effects to be negligible rising through liquid in a tank about 60 cm deep, they found that the bubble surface looks like an umbrella of semi-angle about $50°$ and that although the edge is unsteady in its streamwise extent the upper surface is steady, smooth and accurately spherical. A spherical form near the forward stagnation point on the bubble surface was to be expected, since the variable parts of the dynamic pressure and the hydrostatic pressure at the bubble surface are then both proportional to the square of the distance from the forward stagnation point and evidently cancel at a certain speed of rise. The new conclusion of the investigation was that the observed speed of rise is close to $\frac{2}{3}(gR)^{\frac{1}{2}}$, where R is the radius of curvature of the upper surface, showing that the dynamic pressure variation near the forward stagnation point is close to that which would occur on the forward face of a *complete* sphere moving through frictionless fluid. This result is both interesting and useful, and has been a starting point for many later studies and applications. In a second part of the same paper Taylor and Davies examined the shape and speed of the bubble that runs up an emptying circular tube of radius a which is closed at the top, and found the steady speed of rise to be about $0.48(ga)^{\frac{1}{2}}$ (as Dumitrescu[1] had found in 1943, unknown to Taylor

1. *Z. angew. Math. Mech.* **23**, 1943, p. 139.

and Davies). Note the implication that several emptying-tube bubbles are eventually formed by a large number of spherical-cap bubbles of different sizes in a long vertical tube.

The second postwar paper arising out of work on underwater explosions was on the analysis of the growth of disturbances to a plane interface between two fluids which is accelerating in the direction of the denser fluid (1950f). The linear phase of gravitational instability of an interface had been investigated theoretically by Rayleigh[2] many years previously (unknown to G.I.) and the analysis of instability of an interface in an acceleration-force field is of course mathematically identical. The only actual addition to the theory made by G.I. was to allow for the presence of a second interface, in which case one of the two fluids is initially in the form of a horizontal sheet of uniform thickness. More importantly, G.I.'s analysis was accompanied by some experiments. These were carried out by D.J. Lewis, a research student working under his supervision in the Engineering Laboratory, and, as well as confirming the calculated growth rates of sinusoidal disturbances of small amplitude, they revealed an intriguing finger-like form taken by disturbances of finite amplitude.[3] Ultimately each finger moved steadily into the heavier fluid, like an air interface rising in a vertical tube filled with liquid, thereby establishing an unexpected (or was it?) connection with other work with Davies on rising bubbles. Instability of an interface in a general force field is now commonly referred to as Rayleigh–Taylor instability.

G.I. wrote an interesting paper during World War I on the shape adopted by a parachute during its descent and on the design features needed for stability (1919b). During World War II, when parachutes were used for dropping goods as well as men and porous sheets were used for other aerodynamic purposes such as smoothing the stream in a wind tunnel, G.I. took up work on the relations between the detailed geometry of the strands and apertures of the fabric and the aerodynamic characteristics of the sheet, in particular the local pressure drop across the sheet for a given rate of flow through it and

2. *Scientific Papers, vol II*, pp. 200–7, Cambridge University Press, 1960.
3. *Proc. Roy. Soc. A* **202**, 1950, p. 81.

the resistance to motion of a finite portion of the sheet (1944b, d). I had been concerned myself with the aerodynamic effects of sheets of wire gauze during the war, and in 1947 G.I. and I collaborated in a comprehensive investigation of the effect of a plane sheet of wire gauze on both steady disturbances (independent of position in the stream direction) and unsteady disturbances (turbulence) to the stream of fluid passing through it. The new development that prompted this investigation was G.I.'s realization that as well as causing a drop in pressure a sheet of wire gauze would exert a side-force and so deflect air which passes through it at an angle to the normal. At G.I.'s suggestion, Simmons and Cowdrey at the NPL and Dryden and Schubauer at the National Bureau of Standards in Washington made observations of the amount of this deflection, and it was found that the refraction coefficient was related in a fairly simple way to the pressure drop coefficient. With this information and some new theory for the change in one Fourier component of the velocity disturbance as fluid passes through the sheet, we were able to calculate the smoothing effect of the gauze sheet on steady and unsteady disturbances of small magnitude (1949a). This was the only paper I wrote with G.I. I found the collaboration difficult, because his mind worked much more quickly than mine, and if I had not previously invested a good deal of thought in this problem I should not have been able to retain a sense of making an equal contribution.

The beginning of a Taylor 'school'

The end of the war did not lead to much immediate change in the direction of G.I.'s research, but it did bring about a change in his personal position. The war made enormous demands on the kind of research in mechanical science at which G.I. excelled, and he had become famous (and was knighted and awarded the Copley Medal of the Royal Society in 1944 and awarded the US Medal for Merit in 1946). There was a corresponding increase in the number of people interested in the mechanics of fluids and solids, and the number of research students wishing to work in these fields began to rise. I was among them, and a few words about how I was drawn into contact with the great G.I. Taylor may be of interest.

As an undergraduate at the University of Melbourne I studied mathematics and physics, and enjoyed doing so. I had only a vague notion of what was meant by 'doing research' but I was certain I wanted to learn how to describe physical processes in mathematical terms and to use that knowledge to make discoveries. After graduating in February 1940 I therefore went to see one of the professors in the Department of Physics to seek his advice on what I should do. I was attracted to what was then called 'modern physics', and I asked him whether nuclear physics would be a suitable subject for my proposed research. My advisor said that the war was likely to be a long one, that nuclear physics had no known useful applications, certainly none with any military relevance, and was likely to be a neglected subject for the duration of the war. As things turned out he was quite wrong but what he said was a reasonable conclusion from the public knowledge of that time.

However he also made the constructive suggestion that I should visit the Aeronautical Research Laboratory, which had recently been set up in Melbourne to provide a research and development service for Australia's embryonic aircraft production industry. I did so, and found that one of the staff there (Gordon Patterson, a Canadian) would be happy to supervise the work of an MSc student on aerodynamic problems, the first being the calculation of the corrections needed to allow for the effect of the walls of a wind-tunnel of octagonal cross-section on the measured lift and drag forces on a model. I took up problems of this type and got an MSc for some approximate analysis which I must admit would be regarded as straightforward nowadays. More importantly, since my intended plan of going to England to work for a PhD had to be postponed during the war, I joined the staff of the Aeronautical Research Laboratory. The practical aerodynamic problems that were assigned to me there led me to read about wing and aerofoil theory, boundary layers, shock waves, turbulence, and other areas of fluid mechanics, all quite novel and interesting to me. Gradually the idea of undertaking research in fluid mechanics after the war took root in my mind, and since the aspect about which least seemed to be known was turbulence I decided that would be my PhD topic.

G.I. Taylor at Cambridge was the acknowledged British authority on turbulence, and so when the end of the war was in sight in late 1944 I wrote to ask if he would be willing to supervise my research. Professor Tom Cherry, my teacher at the University of Melbourne, knew G.I. (see chapter 7), and added a note in support of my request. G.I. agreed, and thereby made me a happy man. One of my colleagues at the Aeronautical Research Laboratory, Alan Townsend, intended to return to Cambridge when the war ended in order to complete his graduate studies, and since I knew him to be a first-rate physicist and an electronics wizard I suggested that he too should work on turbulence under G.I. Taylor and that we should collaborate. Being a very adaptable person Alan agreed to do so, and G.I. welcomed the presence of an experimenter, especially one who , like Alan, could cope with hot-wire anemometry. The outcome for both of us was a marvellously exciting decade of turbulence research which began in 1945. Those were halcyon days in which there seemed to be many interesting and useful things to do on turbulence, both experimentally and theoretically, and we had time to do them. Initially we concentrated on the properties of homogeneous turbulence, following G.I.'s pioneering work in the 1930s, but the field of interest was soon widened and Alan in particular made many studies of turbulent shear flows.[4]

As I have already remarked, G.I. himself was not engaged in research on turbulence at this time, because he wished to explore further a number of interesting topics associated with practical problems arising out of war-time defence needs. While always being willing to hear about what we were doing on turbulence he allowed us to go our own way.[5] This suited Townsend and me very well, and I

4. Much of the work in this early postwar period culminated in the books *The Theory of Homogeneous Turbulence*, by G.K. Batchelor, Cambridge University Press, 1953 and *The Structure of Turbulent Shear Flow*, by A.A. Townsend, Cambridge University Press, 1956.
5. Sir Arnold Hall had a similar experience of G.I. as a supervisor in the late thirties. He writes: 'He was immensely encouraging and treated those working under him very much as adult colleagues. Having indicated an area of work he thought would be of value, he left the day-to-day conduct very much to those who were on the job.' Albert Green, another of G.I.'s research students in the late thirties, makes a similar comment:

believe we positively enjoyed and benefited from the need to be independent. Later I observed that while G.I. was always willing to take on a research student for work in an area in which he was interested, he never sought research students and was not personally concerned about educational aspects of their work. But for the self-propelling student who could understand and make good use of G.I.'s insight and illuminating comments, he was ideal.

In 1948 Alan Townsend and I, both of whom were older than the normal research student, were appointed to staff positions at Cambridge, Townsend in the Department of Physics and I in the Faculty of Mathematics. Townsend and I were then formally qualified to supervise research students and to help with the growing numbers of applicants, and from then on there were always several research students at work around G.I. Fenton Pillow (from Australia), Ian Proudman (son of the oceanographer, from England) and Bob Stewart (from Canada) all began in 1948, the first two under my supervision and the third under Alan Townsend's supervision. Turbulence was not the only research area of the new students; but there was a semi-conscious intention to stay within fluid mechanics and to form a group with some cohesion.

G.I. was of course the star of the group, and it was his influence that gave the group its raison d'etre. He was more than 30 years older than every one of us, but was friendly and unassuming although shy and inclined to be a solitary worker. It was his custom to do most of his writing and calculating work (using when necessary a Brunsviga mechanical calculator on long-term loan from the Air Ministry) on a sofa in the drawing room of 'Farmfield' and to come to his room in the Cavendish Laboratory mainly for experimental work. A regular Fluid Mechanics Seminar was initiated in 1948, and has continued ever since. G.I. often attended in the early years but seldom took an

(footnote 5 continued from p.219)
'G.I. always left me to do my own mathematical work, but he could always put his finger on key physical aspects. When I was talking to him he would scribble some ideas on a piece of paper in a very untidy way. I would take this valuable paper away to decipher at leisure.'

active part in the discussions. Visitors came to see him often, and he was generous in giving them his time. Some came to work under him for extended periods, the first in the postwar years being Les Kovasznay, a Hungarian, who held a British Council Scholarship at Cambridge during 1946.

In these early postwar years there was a severe shortage of working space. G.I. had at his disposal two rooms in the Cavendish Laboratory, one of which was almost completely filled with the low-turbulence wind tunnel built by D.C. MacPhail and used by Alan Townsend in his experiments. Desks were fitted in wherever there was sufficient space, and the Cavendish Professor (at that time Sir Lawrence Bragg) generously turned a blind eye to whether the occupants were attached formally to the Department of Physics or to the Faculty of Mathematics (which, incredible though it may seem, had no working facilities of any kind). We learnt by experience that a connection with Professor Sir Geoffrey Taylor was valuable.

CHAPTER 16

The second golden period 1951–72

Retirement from the Yarrow Research Professorship

In 1951 G.I. reached the age of retirement from his Yarrow Research Professorship. His salary was reduced, but the change in formal position made absolutely no difference to what he did or to the way he did it; indeed, there were many in Cambridge who were unaware that he had 'retired'. The important point was that his room in the Cavendish Laboratory and the services of his assistant Walter Thompson continued to be made available to him. These were the realities of life for him.

By a coincidence the nature of his scientific interests changed at about the time of retirement from his professorship, evidently because he had by then carried to completion the many investigations that arose out of war-time needs. There began, around 1951, a new phase of his scientific life which was to last about 20 years and in which he explored a remarkable range of novel and unconventional problems in fluid mechanics. Whereas in earlier years his choice of scientific problem was determined mainly by external factors and influences, such as his membership of the Aeronautical Research Committee, in this later phase I think his choice was more deliberate and reflected his personal preferences. In these postwar years scientific research was becoming popular, well supported and specialized, and there were many people working in the fields regarded as important. As a confirmed individualist, G.I. found much more fun in the problems that others had not yet recognized as being significant. But whereas in prewar years he quickly took the lead in many fields and seemed to have them almost to himself, it was not so easy now. He began to be aware that the development of many new experimental and mathematical techniques was putting him at some disadvantage for certain kinds of research. It was his habit to use simple methods, but it was becoming more difficult to accomplish with simple methods what

Figure 16.1 *G.I. in 'retirement' in 1955 (age 69), in the Cavendish Laboratory with his assistant Walter Thompson.*

others were doing with sophisticated techniques. For these reasons I believe that after 1951 he deliberately sought problems in fluid mechanics which lay outside the main streams of research and which could be explored by simple methods; and of course he continued to be specially attracted to problems which allowed both theoretical and small-scale experimental investigation.

It was during this period that G.I. began to refer to himself as, to some extent, a scientific amateur, like his grandfather George Boole, his aunt Alice Boole, and his grandmother's uncle George Everest (all of whom were introduced in chapter 2). This has puzzled many people. In 1962 G.I. gave an address on the occasion of the Semicentennial Anniversary of Rice University (1963c), mostly about the scientific work of the famous members of his family and others who despite being untrained made useful contributions to science with simple apparatus and ideas. His initial intention was to entitle the lecture 'Amateur scientists', but on consulting the dictionary and finding that the word 'amateur' may be used to describe work which is not up to professional standards, he saw that this was not quite his meaning and changed the title to 'Scientific diversions'.[1] HowHowever, this does not clarify the matter completely. Bearing in mind what he achieved with simple methods, the humility of the following remark in the final paragraph of the text of his address at Rice University is astonishing: 'In this talk I have tried to convey the idea that the pleasure and interest of being a scientist need not be confined to those gifted people who have the ability to pursue the highly specialized studies which are necessary for those who would reach the main frontiers of scientific advance.'

My 'Unfinished dialogue with G.I. Taylor'[2] gave me the opportunity to put questions to G.I. directly, and I asked him what he had in mind when referring to himself as in some sense an amateur in science. G.I.'s reply began with a description of his mathematical training at Cambridge, and then went on as follows: 'Though the mathematical

1. But a few years later, on a similar occasion at the University of Michigan, he gave an address containing much the same material with the original title (1969d).
2. *J. Fluid Mech.* **70**, 1975, pp. 625–38.

methods I have been taught have proved adequate for many of the physical problems I have studied, they are now regarded as old fashioned and I have not familiarized myself with the modern notations. In that sense I am like an amateur who takes up a subject and works on it without intensive training, using mainly instinctive reasoning. I know that the word amateur is often used in a pejorative sense but I think of it as meaning a person who does something because he wants to, even if he has not been intensively taught how. In that respect I feel I follow at a great distance my grandfather, George Boole, and others like Benjamin Franklin and Ramanujan. I do not really refer to myself as an amateur but only as one who, like an amateur, has not mastered the modern techniques for doing his work'.

It appears therefore that in these later years of his life G.I. felt handicapped by his lack of familiarity with modern mathematical and experimental techniques, but was encouraged by the success of his famous grandfather and aunt, who experienced more severe handicaps of this kind, and was resolved nevertheless to enjoy the pursuit of science like his relatives and the well known 'amateurs' featured in his Rice University address.

A feature of G.I.'s post-retirement years was the further growth of the group of young people working around him in the Cavendish Laboratory. He was not involved closely with much of this work in fluid mechanics but his presence was a great stimulus. To an increasing extent he sought help from young colleagues who had useful expertise in some technique unfamiliar to him – electronics, numerical computations, analysis – although the collaboration did not often go beyond use of the expertise. In 1955 someone had the bright idea of taking the group photograph reproduced in figure 16.2 – a pity for biographers that it did not become an annual event. At this size the group was acutely short of accommodation, but the Cavendish Professor, Sir Lawrence Bragg, continued to allow the group to occupy odd corners of space. A little later a properly equipped room was provided for G.I. and his mechanic Walter Thompson and the wind-tunnel constructed by Donald MacPhail and G.I.'s collection of apparatus, and again some of the research students engaged in theoretical work could find a vacant corner for their desks.

Figure 16.2 *The Fluid Dynamics Group in the Cavendish Laboratory, April 1955.*
FRONT ROW: *Tom Ellison (Res. Fellow), Alan Townsend (Assis. Dir. Res. in Physics), Sir Geoffrey Taylor, George Batchelor (Univ. Lect. in Math.) Fritz Ursell (Univ. Lect. in Math.), Milton Van Dyke (Visitor).*
MIDDLE ROW: *S.N. Barua, David Thomas, Bruce Morton, Walter Thompson, Owen Phillips, Freddie Bartholomeusz, Roger Thorne (all research students except Thompson, G.I.'s technician).*
BACK ROW: *Ian Nisbet, Harold Grant, Anne Hawk, Philip Saffman, Bill Wood, Vivian Hutson, Stewart Turner (all research students).*

The lack of office facilities for the staff of the Faculty of Mathematics was ultimately recognized by the University, and in 1959 a Department of Applied Mathematics and Theoretical Physics was established, within the Faculty of Mathematics, in order to provide a formal framework for the acquisition and use of buildings and facilities and secretarial staff. It was a big step forward, but it was not until 1964 that an old building near the centre of Cambridge (formerly occupied by Cambridge University Press) was acquired by the University and converted for use by the new Department. This building yielded an ample number of offices on three floors, and in the basement there was space for a small laboratory suitable for experimental work on fluid mechanics. There was some opposition from the Department of Physics to the idea of spending money on apparatus for use by untrained theorists in the new Department, but the University accepted the claim that research in subjects like fluid mechanics flourished when theorists and experimenters could work together. The DAMTP laboratory was thus established, and has had a valuable influence on the whole programme of research in fluid mechanics in the Department. Experimental work in the laboratory was under the general direction of Brooke Benjamin initially, then Stewart Turner, and currently Paul Linden. Understandably, at the age of 78, G.I. preferred to avoid the upheaval of another move and continued to do his experimental work in the Cavendish Laboratory, coming to the new Department occasionally for seminars.

During the years of his retirement G.I. travelled abroad to many conferences and meetings of different kinds. He had always liked foreign travel and seeing new places, and now that he was a great man of his subject and on everyone's list of invited participants he had many opportunities; and the further and more unfamiliar the location the better. He enjoyed being feted at these meetings, being made 'honorary president' or presented with a medal, with the simple pleasure of one who never took it for granted that his work deserved recognition. Very often the organizers of a meeting expected this great authority to give a lecture on a topical subject and to make pronouncements on current research; instead there would be a refreshingly modest and enthusiastic (and sometimes incoherent) description of some neat little experiment on jets or waves on a sheet of liquid or peeling off a strip of adhesive, not always connected very closely with the theme of the meeting. He was particularly good with young people at international meetings, being approachable and friendly and so obviously not an organization man, and there are many who are proud to remember that they entertained him in their homes and that he enjoyed it.

It was natural that G.I. should occasionally be asked to address a general audience, perhaps on the occasion of conferment of some academic honour. He usually accepted such invitations, and, although he was not impressive as a platform speaker, the accompanying published text disclosed a charming style of writing, precise, light and modest, with a gift for unsophisticated reminiscences and amusing observations on the (scientific) life that he had led and the scientific fields with which he had been involved. Some of these general articles were about George Boole and his family (1954f, 1956e, 1964c), some were memories of scientific contemporaries (1953e, 1962c, 1963b, 1963d, 1973b), some were reflections on the history of scientific developments in which he had taken a leading part (1957b, 1965b, 1966f, 1969e, 1970b, 1971b, 1971c), and others – the most interesting and revealing ones – were about his life and work (1952d, 1956f, 1963c, 1969d, 1970a). I have learnt a good deal about G.I.'s opinions and attitudes and values from these articles, possibly more than I learnt from personal contact with him over a period of 30 years. As a small, previously unpublished, sample of these delightful articles, I am reproducing in Appendix A his (incomplete) address of reply, entitled 'An Applied Mathematician's Apology', after receiving the de Morgan medal from the London Mathematical Society in 1956.

G.I. was eminently successful in finding research topics which met the above-mentioned requirements of his retirement years, and I can identify five or six quite new and interesting fields in each of which he wrote several papers and which subsequently became popular when others also saw their potentialities. In the 21 years beginning in 1951, that is, between the ages of 65 and 86, he wrote 48 scientific papers, most of which are substantial and some of which are pure scientific delight. The total achievement of these 21 years would alone be more than enough to ensure the outstanding success of a scientific career. He retained in these retirement years his characteristic freshness of approach and delight in new phenomena, and advancing age seemed to make little difference to his insight and capacity for original thought. G.I.'s papers in this final phase of his working life also show, perhaps more clearly than those in any other period, his uncanny ability to select significant problems. I find it difficult to put my finger on exactly what constitutes significance, but it implies a fundamental quality and a relevance or applicability to a variety of contexts. Somehow G.I. was able to recognize significance intuitively, and many of the papers he wrote in these years (for example, those on longitudinal dispersion in flow through tubes described later in this chapter) have been the starting point for numerous further developments. His skill as an experimenter was never more evident, and in my view some of these later papers are more beautiful gems of scientific investigation than those on larger themes written in his more vigorous years. The post-retirement years 1951–72 constituted a second golden period of G.I.'s research which was perhaps not as grand as the first but was undoubtedly more fun.

An account of G.I.'s scientific work during the period 1951–72 inevitably looks like a catalogue, since he investigated so many different types of problem. A few words about each of the main new areas of research must suffice. Before turning to them I should like to mention a short paper which does not qualify as a new area of research since it probably did not occupy more than an hour or two of his time but which is a perfect example of his style of thinking. In April 1954 the Royal Society held a discussion on 'The first and second viscosities of fluids', in which G.I. took part. The second coefficient of viscosity (often termed the expansion viscosity) was evidently a rather mysterious concept then, and the record of the discussion is full of learned contributions from the thermodynamical, rheological and molecular points of view. G.I. always found it helpful to have a concrete picture, and so he conceived a fluid medium for which the effective expansion viscosity could be calculated explicitly, namely an incompressible liquid containing some well separated small spherical gas bubbles. The way in which dissipation occurs during expansion of this medium is absolutely clear, being a consequence of the shear viscosity of the liquid and the radial motion of the liquid near each bubble, and the expansion viscosity may be found from a few lines of working (1954d). This specific case shows that the expansion viscosity is associated with the lag in the adjustment of the pressure in the gas bubbles to its equilibrium value while the medium is expanding – and this is clearly a fundamental way of thinking about the expansion viscosity generally. There is no longer any mystery. I doubt if any of the general arguments and conclusions presented at the discussion are remembered today, but G.I.'s deceptively simple little note has provided a textbook example – still the only one available – of a real medium with a calculable expansion viscosity.

The swimming of small creatures

The first new topic to appear in G.I.'s papers after World War II was the mechanics of swimming of microscopic organisms. G.I. says he was introduced to this topic by Sir James Gray and Lord Rothschild, and I remember Rothschild taking G.I. and me to the University Farm to show us fresh bull spermatozoa under a microscope. G.I. was intrigued by the novel mechanical feature of the problem that inertia effects are negligible in motion on such a small scale and that viscous forces are dominant. Thus the generation of thrust by the usual process of inertial reaction is impossible; or, as G.I. put it, the progress of a microscopic organism more nearly resembles a screw going into wood than a fish swimming. He correctly perceived that the study of microscopic swimming creatures opened up a new field in hydrodynamics – a 'virgin' field, as he daringly described it in a seminar. In his first paper (1951a) he considered the simplest possible model of a wriggling self-propelled organism, namely an infinite flexible but inextensible sheet on which waves of lateral displacement are propagated with speed V, and he was able to show that the total hydrodynamic force on the sheets is indeed zero if it moves in the direction opposite to that of the wave propagation, with a certain speed which is proportional to V and to the square of the ratio of the wave amplitude to the wavelength (this ratio being assumed small in the analysis). A striking feature of the motion of spermatozoa is that when two heads happen to be close together the tails beat in unison, with the distance between adjoining parts of the two tails remaining roughly constant, and this prompted G.I. to show from his model that the phase of the hydrodynamic stress at one of two parallel sheets is such as to bring the phases of the displacement of adjoining parts of the two sheets closer together.

Having shown that propulsion by the propagation of lateral waves is possible in principle, G.I. sought quantitative results for a flagellum by representing it as a flexible circular cylinder of infinite length on which waves of lateral displacement are propagated (1952b). He was able again to calculate the hydrodynamic stress at the surface, correct to the second order in the wave amplitude, and to show that the total hydrodynamic force vanishes if the cylinder is moving at a certain speed parallel to its length, both for a plane wave and for a wave of helical form. (It is known now, but was not then, that some flagella beat with plane waves and some with helical waves.) The organism is propelled forward essentially because every element of the cylinder behaves approximately like an obliquely moving rod; and since lateral resistance to motion of the rod is greater than longitudinal resistance (as may be shown by slender-body theory), lateral motion of each element generates a resultant longitudinal component. The analysis showed also that a helical wave generates a circulatory fluid motion in a transverse plane, implying that a torque must be exerted on the cylinder about the axis of the helix if it is to be prevented from rotating as a whole, as Gray had previously concluded.

Some years after publication of this work on the way in which microscopic organisms can swim in spite of the smallness of the Reynolds number of the flow, G.I. accepted an invitation from Educational Services Inc. to direct the production of an educational film on 'Low-Reynolds-number flows'(1967a). This organization had already produced several films of high quality on different aspects of fluid mechanics, on the initiative of Professor A.H. Shapiro at Massachusetts Institute of Technology, and the choice of G.I. as director of a new film was inspired. This role gave him scope for his unmatched ingenuity in devising laboratory experiments to illustrate the principles of flow with small inertia forces. Among the demonstrations devised for the film was a comparison of the swimming speeds of two models, first in water and then in viscous syrup. One of these models has a tail-plate which is made to oscillate by means of a twisted rubber band and swims well in water by inertial reaction, but not in syrup (remember the reversibility of low-Reynolds-number flow). The other consists of a section of a wire helix which is made to rotate about its axis by a twisted rubber band, and swims forward in the viscous syrup but makes no progress in water. Figure 16.3 shows a few still-photographs taken from the film, which is full of similarly ingenious demonstrations and is still of the greatest education value.

In a third paper on the mechanics of swimming G.I. turned his attention to larger long narrow animals such as snakes, eels and marine worms which also swim by propagating waves along their length (1952c). Inertial forces are relevant here and exact mathematical analysis of the fluid motion is not feasible, so G.I. assumed that the

(a) (b)

(c)

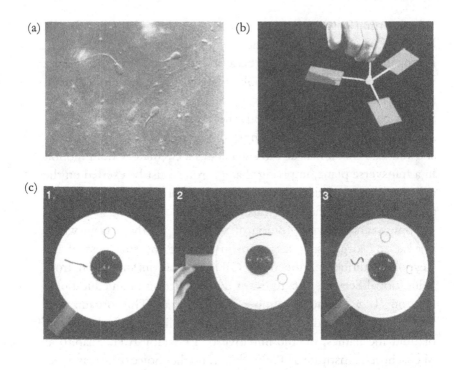

Figure 16.3 *Still photographs from the film Low-Reynolds-Number Flows (1967a).*
(a) Bull spermatozoa, about 0.05 mm in length. (b) The teetotum, a toy to demonstrate
hydrodynamic lubrication. The three lamina are each inclined slightly, and when
spun counter-clockwise on a table the front edges are higher. A high pressure which can
support the weight of the teetotum is then generated in the fluid layer. (c) A rigid ring, a
piece of wool, and a square blob of dye are inserted into glycerine in the space between two
circular cylinders initially at rest (c1). The inner cylinder is then turned N times in one
direction (c2) and N times in the other direction (c3). The dye, remarkably, returns to
its initial compact form, showing the reversibility of the flow. Only the motion of the
flexible wool is not reversible.

hydrodynamic force exerted on an element of length of the deformed
cylindrical animal at any instant is the same as if that element were
part of a straight circular cylinder of infinite length having the same
velocity as the element (as he previously did in the analogous problem
of determining the shape of a balloon cable in a wind). Data concerning
the transverse component of force acting on unit length of rough-
and smooth-surfaced circular cylinders set obliquely to the wind were
available from wind-tunnel measurements; and G.I. obtained plausible

estimates of the longitudinal component by reinterpreting measurements of heat transfer from cylinders. The resultant force acting on a flexible cylinder on which plane waves of lateral displacement are propagating was then calculated, and the combination of forward speed, wave frequency, length and amplitude which would make this force zero was determined. There are too many parameters for the results to be striking or quotable, but one surprising prediction was that a long narrow animal with a very rough surface should be able to propel itself by propagating waves in the *forward* direction. On hearing about this, Gray showed G.I. a marine worm with flexible projections on its surface which does in fact swim in this way.

Longitudinal dispersion in flow in tubes

G.I. records that the idea for his next new research topic came when a young physiologist asked for help in understanding how drugs are dispersed when injected into blood vessels of animals. He apparently did not do any work having a direct physiological application, but through thinking about the dispersal of material injected into a cylindrical tube containing liquid in motion he was led to some very important basic results which have been used in many different contexts. Taking first the case of steady laminar flow in a circular tube, he showed, by rather clumsy analysis which does not make the best starting point for a student, that at a large distance downstream from the injection point the concentration of material is approximately uniform over the tube cross-section and the longitudinal spreading is governed by the diffusion equation (1953c, 1953d). Thus the centre of a blob of injected dye or salt in solution ultimately travels downstream with the mean speed of the fluid U_0 and its width in the flow direction increases as the square root of the distance travelled; fast-moving fluid at the centre of the tube which catches up with the blob and acquires material by molecular diffusion evidently loses it to the slow-moving fluid near the wall when it moves ahead of the blob. The diffusivity for this longitudinal dispersion process was found to be $a^2 U_0^2 / 48D$, where a is the tube radius and D is the molecular diffusivity of the material. This expression has the interesting property of being larger for smaller values of D, because the differential longitudinal convection is less hindered by lateral molecular transfer when D is smaller.

G.I. made observations, by a colorimetric method, of the average concentration of potassium permanganate over the cross-section of a tube at different distances downstream from an injection point, and found, as predicted, that an injected blob ultimately has a Gaussian distribution about a point at distance $U_0 t$ downstream. Confirming the theoretical expression for the diffusivity was more difficult, because accurate values of the (concentration-dependent) molecular diffusivity of the dye were not available; indeed G.I. suggested that observations of longitudinal dispersion in a tube might provide the best method of measuring D (1954c).

The flow of liquid in a tube is likely to be turbulent except when the tube diameter is quite small, and G.I. saw the practical importance of extending these considerations of longitudinal dispersion to turbulent flow. Water engineers had found empirically that the position of maximum concentration of salt injected into a water main arrived at some point far downstream at a time which corresponded with travel at the mean water speed, and observations of the longitudinal dispersion of radioactive tracer elements in long oil pipelines had suggested a Gaussian distribution of concentration, but there was no theory which would allow interpretation and use of the observations. G.I. perceived the essential similarity of the cases of laminar and turbulent flow, with the molecular diffusivity in the former case being replaced by an eddy diffusivity in the latter, but the problem was to relate this eddy diffusivity, which is an imprecise concept, to the measurable properties of the turbulent flow. He carried through the calculation of the diffusivity for the longitudinal dispersion, by a method involving use of the Reynolds analogy between the lateral transport of heat and momentum in a turbulent flow (1954a). I think only G.I. could have handled with confidence this mixture of analysis and empirical information about turbulence and emerged with a reliable answer. With the (considerable) help of Tom Ellison, a Research Fellow of Clare College, G.I. also made conductivity measurements of the concentration of salt at a fixed station at different instants after injection at a point far upstream in a pipe of 1 cm diameter and found a longitudinal diffusivity close to the calculated value $10au_*$ for both smooth and rough pipe walls, where u_* is the 'friction velocity'.

This investigation of longitudinal dispersion raises interesting questions concerning G.I.'s research tactics. He must have realised from the start that, if any interesting properties of the dispersion process exist, they will apply asymptotically far downstream from the point of injection of dye. It is natural then for an applied mathematician to see a statistical role for the lateral diffusion, and to anticipate a possible application of the Central Limit Theorem. G.I. makes no mention of such an application, but it can in fact be seen readily from this theorem that the probability density function of the downstream displacement of a dye molecule tends to the Gaussian form as the mean displacement tends to infinity (think of the total displacement as the sum of a number of displacements due to convection in equal time intervals). Thus the longitudinal dispersion is asymptotically a diffusion process whenever the velocity of a dye molecule is a stationary random function of time (for only then do the displacements in the different time intervals all have the same mean square), which includes the cases of steady laminar or turbulent flow in straight tubes of arbitrary cross-section, or in regularly corrugated or curved tubes. But powerful though this general result may be, it does not give the numerical coefficient in the expression for the effective longitudinal diffusivity. This coefficient, which is needed for a comparison of the theory with observation – and which would consequently be regarded by G.I. as the most important product of the investigation – depends on the distribution of the fluid velocity over the tube cross-section and so requires a separate calculation in each case. G.I. always took a practical view of a problem and attached relatively little value to elegant general results if they did not permit a comparison with observation.

Movement of an interface through narrow passages

On passing over an interesting little group of papers on flow in regions bounded by porous surfaces and the draining of water from a sheet of paper pulp in a paper-making machine (1956b, 1956c, 1957b), the next broad new field to be taken up by G.I. can be described vaguely as being concerned with the way in which an interface between two fluids moves through narrow passages of different kinds. In G.I.'s hands this rather esoteric topic led to a remarkable number of practical problems and useful results. G.I. had been struck by the simple finger-like shape of the fully developed unstable disturbance to an accelerating interface between two inviscid fluids found by Lewis,[3] and he thought it likely that something similar might occur in other situations. Stimulated by a visit to the Humble Oil Company in Houston, he noticed theoretically that if a plane interface between two fluids moves normally through a porous medium, the interface is unstable to all small disturbances, in the absence of effects of gravity, when the interface advances toward the more viscous fluid; and his intuitive expectation that finger-like disturbances of finite amplitude would be formed in this case was confirmed by some experiments, not with a true porous medium but with a Hele–Shaw cell[4] (see figure 16.4). With Philip Saffman he examined analytically and experimentally the shapes of such fingers advancing steadily through the cell (and also those of bubbles of finite volume) and found an infinite family of shapes with different values of the ratio of finger width to finger spacing (1958b, 1959a). The experiments showed that this ratio seemed always to be a half, except at small interface speeds when the effect of surface tension at the interface was important, but Tayor and Saffman were not able to advance any physical reason for the occurrence in practice of this particular member of the mathematical family of shapes. The point is of some importance, because the ratio of finger width to spacing determines the fraction of oil which may be driven out of a porous bed by an advancing water interface. Explanation of the mathematical indeterminacy came many years later and was found to involve the effect of surface tension at the interface between the two fluids in the Hele–Shaw cell in a complex way which makes an intriguing story[5] too long for inclusion here. Once again G.I. had spotted a phenomenon of significance.

3. *Proc. Roy. Soc. A* **202**, 1950, p. 81.
4. A thin layer of fluid, between two parallel plates, whose motion parallel to the plates is governed by the two-dimensional form of the Darcy equation.
5. The definitive reference is, very appropriately, by Philip Saffman ('Viscous fingering in Hele–Shaw cells', *J. Fluid Mech.* **173**, 1986, pp. 173–94) who collaborated with G.I. in several investigations.

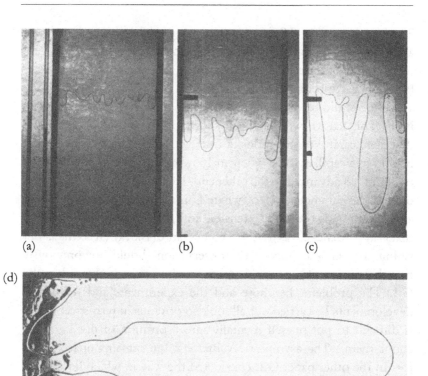

(a)   (b)   (c)

(d)

(e)

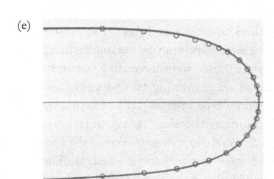

Figure 16.4 *Stages in the development of an initially straight air–glycerine interface as it is pushed through a Hele–Shaw cell. The forward portion of the finger in (d) is compared in (e) with the points found theoretically for a finger to spacing ratio 0.5. (From 1958b)*

The fact that a second fluid advancing into a Hele–Shaw cell does not always displace completely the first fluid and may leave some adhering to each plate of the cell (although without affecting the shape of interfacial fingers) prompted G.I. to consider flow systems in which the amount of fluid retained at a rigid surface by friction is of interest in itself. This led to two delightful short papers. In one of them G.I. described careful observations of the fraction m of viscous liquid left behind in a capillary tube when air blown into one end of a tube full of liquid has advanced to the other end. He found that m is a function only of the parameter $\mu U/\gamma$, where U is the speed of the advancing air column, γ is the surface tension and μ the liquid viscosity, and that m approaches 0.6 as $\mu U/\gamma$ becomes large (1961a). (It seems astonishing that such a simple basic observation should not previously have been made. One has this reaction often to something done by G.I. The problems he chose and the experiments and theoretical developments he carried out all look so obvious in retrospect that it is difficult to put oneself mentally into a position of not knowing about them.) The asymptotic value $m = 0.6$ remains unaccounted for. In the other paper G.I. considered the way in which liquid may be deposited on a plane rigid surface, either by a model paint brush consisting of a set of closely spaced parallel plates normal to the surface over which they are sliding or by a liquid-filled porous roller (1960b). In both cases some simple theory gave an estimate of the thickness of the layer of liquid left on the surface which was in agreement with his measurements.

G.I. noted that the conditions immediately behind the porous roller, where fluid divides into a layer remaining on the roller surface and a layer left on the plane surface, raised some interesting questions and he returned later to them in the remaining two papers of the group. The first is a long and mainly experimental study of cavitation bubbles in a hydrodynamically lubricated bearing (1963a), and is notable for the convergence of the lines of thought in several of the previous papers on movement of interfaces in narrow passages. The main new finding of this relatively little known paper is that two physically different types of cavitation may occur, one being the conventional type which depends on the occurrence of a sufficiently low pressure for gas to come out of solution, and the other being akin to the advancing meniscus of a gas bubble in a tube or to the meniscus behind a porous roller spreading liquid on a plane surface and, like these minisci, influenced by surface tension. The second of these two remaining papers is an investigation, with Angus McEwan, of the way in which a flexible strip stuck to a rigid plane surface with a liquid adhesive peels off it when one end is lifted (1966e). G.I. saw that the understanding he had obtained of the air meniscus advancing into the liquid-filled crevice between two separating surfaces enabled him to construct a viscous liquid model of a peeling adhesive, different from the previous models which had assumed some elasticity of the strip. Taylor and McEwan determined theoretically the velocity and stress distributions ahead of the advancing meniscus, and made model experiments which among other things showed the now familiar formation of fingers on the advancing interface (the appearance of which was noted to be similar to that of the zone of rupture of commercial pressure-sensitive tapes). The theoretical relation between peeling force and the (constant) peeling angle, which involves the surface tension except at large values of $\mu U/\gamma$, was confirmed by measurements with the model.

The dynamics of thin sheets of liquid

This was a fascinating study of kitchen-sink phenomena which provided scope for G.I.'s experimental ingenuity and yielded many beautiful photographs and intriguing results. He published three papers together (1959b, c, d), of which the first is about the shapes of the 'water bells' formed by placing a conical obstruction in the path of a narrow jet of water (figure 16.5). The second is an investigation of the two kinds of surface-tension-controlled wave which may propagate on a sheet of liquid of thickness h. Antisymmetrical waves displace the two surfaces of the sheet equally (like a flapping flag) and are non-dispersive, having a wave speed proportional to $h^{-\frac{1}{2}}$. Optical reflexion pictures confirmed G.I.'s prediction that the wavefront of a stationary disturbance formed by a small obstruction at one point of a sheet of uniform thickness and velocity is in the form of two narrow line-like waves whereas the wavefront in the case of a radially expanding sheet is a cardioid (figure 16.6). Symmetrical waves, on

(a)

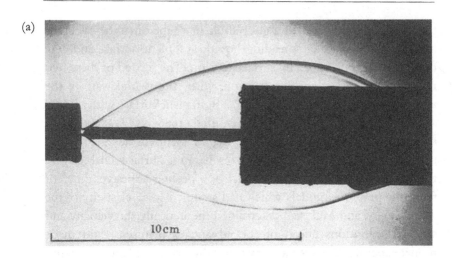

10 cm

(b)

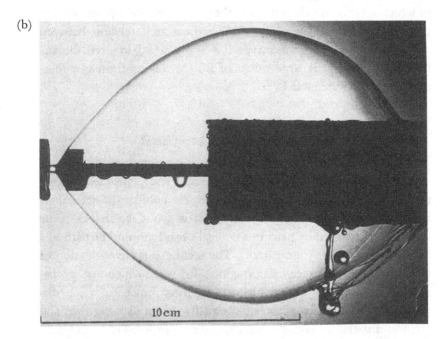

10 cm

Figure 16.5 *Water bell formed by directing a cylindrical jet of water at a conical impacter. (From 1959 b)*

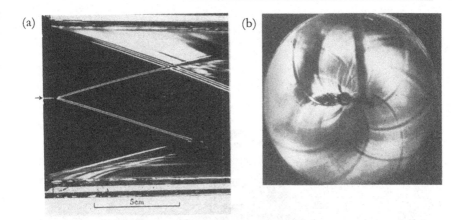

Figure 16.6 *Antisymmetrical non-dispersive waves; (a) on a sheet of water of uniform thickness and velocity, (b) on a radially expanding sheet. In case (b) the wavefront of each of the eight stationary disturbances is a cardioid. (From 1959 c)*

the other hand, cause thickening and thinning of the sheet and are dispersive, shorter waves propagating more quickly, and the wavefronts from a stationary source, which here need to be photographed by transmitted light in the Schlieren manner, are quite different, being parabolas in the case of a moving sheet of uniform thickness.

The third paper of the series concerns the disintegration of the free edge of a thin liquid sheet. The free edge of a sheet is pulled into the sheet by surface tension, at a speed which G.I. showed to be the same as the speed of antisymmetric waves, in the case of a sheet of uniform thickness and velocity. The two free edges stemming from a hole made by a stationary disturbance to a moving uniform sheet are thus straight, although the corresponding edges in the case of a radially expanding sheet are not cardioids (see figure 16.7). The paper included an investigation of the size of the drops produced by a disintegration at the edge of the sheet and the action of swirl atomizers, a topic on which G.I. had written previously (1948c, 1950e).

A little later G.I. took up the problem of the formation of a plane liquid sheet by the oblique impact of two cylindrical jets (1960a). The quantities of interest here are the distribution of thickness of the sheet with respect to azimuthal angle about the impact point and the shape of the free edge of the sheet where it breaks up into drops.

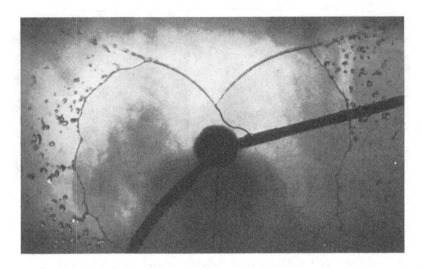

Figure 16.7 *Disintegration of the free edges of a radially expanding sheet separated into two sheets by a stationary obstacle. (From 1959d)*

G.I. measured both, and showed that they were related in the manner expected.

He returned to the topic of thin sheets of liquid some years later, in an intriguing paper with Hugh Michael (one of the earliest PhD students in the postwar Taylor group) on the equilibrium and stability of holes made in a liquid sheet in different ways (1973a). Observation of the impact of small particles on a freely falling thin sheet of liquid in air suggested to G.I. that large holes in the sheet expand whereas small holes close up. This is what one would expect from the rough argument that the surface area of a sheet of uniform thickness t with a cylindrical hole of radius a decreases both when a hole such that $a < t$ closes up and a hole such that $a > t$ opens out. However, G.I. sought more definite results. He noticed that the two principal curvatures of a soap film extending between two co-axial rings are equal and opposite, like those for an axisymmetric hole in a thin sheet of liquid, and he used this analogy to show that there can be no equilibrium form of a hole in a thin sheet of liquid under the action of surface tension alone. On the other hand it was known that a hole in a thin sheet of liquid at rest on a horizontal table can exist in equilibrium under the action of a surface tension *and* gravity. Some nice experiments showed that the equilibrium form of a hole in a horizontal sheet of mercury on a table under water is unstable, with larger holes opening further and smaller holes closing up. Another case studied was the behaviour of a hole in a sheet of water resting on a horizontal sheet of paraffin wax, the holes here being made by a jet of air. G.I. found in this case that holes of different initial sizes remained stationary. This was shown to be associated with the fact that the angles of contact with the paraffin wax were different for advancing and receding menisci.

Electrohydrodynamics

This fifth and last of the major new themes in G.I.'s retirement years was not entirely new to him since many years earlier he had written a short note with C.T.R. Wilson on the shape of an uncharged soap bubble in a uniform electric field and the approach to bursting as the field was intensified (1925c). He wrote six substantial papers on electrohydrodynamics in his retirement years, the first being published when he was 78, and the last when he was 83. I remember that when the last of these six papers was complete he told me that he had written his last paper and that it was time to publish the fourth and final volume of his *Scientific Papers*. It was effectively his last word on electrohydrodynamics, but, as I half expected would happen, he wrote a couple more papers on other topics (1971a, 1973a) after vol. IV had been published.

In the early work with Wilson, G.I. had noticed that when the electric field strength reached a critical value a fine jet issued from the end of the distorted soap bubble; and others had observed such jets issuing from liquid drops. G.I. now examined more closely this interesting phenomenon, which has a bearing on the disintegration of water drops in thunderstorms, and unravelled the underlying mechanics (1964a, 1966a). He showed, by means of a characteristic combination of delicate experiments on water drops in an electric field, and analysis which went no further than necessary, that the discharge of a jet (which might in reality be pulsed) is associated with instability of the drop surface, that as the electric field strength approaches the critical value the end of the distorted drop becomes a cone of semi-vertical angle 49.3° (a by-product of a nice local solution for a conical interface), and that at field strengths slightly in excess of the critical value the instability takes the form of a violent in and out movement of the extreme tip of the cone with a jet being flung off during part of each cycle.

The possibility of jets of conducting liquid moving under the influence of an electric field evidently took G.I.'s fancy, and as a preliminary to the study of jets he considered the force exerted on a long conducting rod resting normally on a conducting plane by an electric field normal to the plane. With the aid of slender-body theory like that developed for hydrodynamic problems, the distribution of induced charge along the rod and the total force on it could be calculated, and experiments showed that light rods were lifted off the plane at the field strength for which the calculated force was equal to the weight. Heavier rods, on the other hand, either oscillated or did not move, evidently because there was electrical breakdown of the air near the upper end of the rod where the field was intense (1966b). G.I. next examined the jets drawn from the free surface of a conducting liquid at the open end of a vertical tube by a potential difference between the liquid and a horizontal plate above the liquid. The free surface likewise becomes conical as the potential difference is increased, and ultimately discharges a jet, but here it was found that the surface shape and flow into the jet were steady. This led naturally to an investigation of the stability of electrically driven jets in different circumstances. A variety of forms of instability was found, one of which is like the crumpling of an elastic rod subjected to end compression, but analysis of the effect of induced charges on the jet surface did not account for these forms completely (1969a, 1969c).

If jets could be drawn from an initially spherical surface of conducting liquid by an electric field, they might also be drawn from an initially plane surface, and so G.I. examined (with Angus McEwan) the stability of a plane interface between a conducting fluid and a non-conducting fluid in the presence of a normal electric field. They showed theoretically that when the electric field strength exceeds a certain value, which depends on the surface tension and density difference, the interface is unstable to small disturbances characterized by a certain length scale. Their experiments with several pairs of fluids confirmed the predicted instability, and showed in particular that an air/water interface is unstable in a way which leads to fine jets of water being ejected towards the upper electrode at field strengths smaller than those at which sparking occurs by electrical breakdown of the air (1965a).

The remaining two papers on electrohydrodynamics were concerned with drops of liquid in the presence of an electric field, and were sequels to the initial study of the deformation of a conducting drop. Previous observations of the shape adopted by a non-conducting dielectric drop suspended in another liquid had shown that both prolate and oblate spheroidal forms were possible, and attempts had been made to analyse these shapes on the assumptions that the interface is uncharged and the liquids are everywhere stationary. G.I. perceived that, however small the liquid conductivities might be, a steady state does not occur until a surface charge has accumulated and that a circulatory motion must exist in both liquids. The drop shape and the circulatory motions in the steady state depend on the ratios of the viscosities, conductivities and dielectric constants of the two liquids and G.I. showed that for a certain combination of these ratios the drop is spherical. Criteria for the drop to be of prolate or oblate form and for the direction of the internal circulation were also derived and found to be consistent with the available data (1966c). In the second paper on drops G.I. examined the coalescence of two neighbouring liquid surfaces held at different electric potentials. Some simple analysis and some equally simple experiments (both of which would be within the capacity of a student, once the objective had been formulated) showed that two soap films stretched between parallel circular rings remain separated only when the potential difference is less than a certain critical value and that there is a discontinuous change in their shape when this value is reached. Further and more difficult experiments established the existence of a similar critical potential difference for two neighbouring spherical liquid surfaces anchored to rings, the critical value being proportional to the spacing at zero potential difference (1968a). It was thought that each surface might take the familiar conical form just before the two surfaces sprang together, but this could not be detected.

CHAPTER 17

The closing years

T he serenity of G.I.'s features in his eighties testified to his stress-free life. The photograph reproduced in figure 17.1 was taken by Angus McEwan, one of his former research students, in 1971, when G.I. was 85, and one may see there the peace and composure that he had enjoyed throughout his life. But that photograph proved to be the last record of G.I. in a well state.

In April 1972 G.I. suffered a severe stroke which paralysed most of the left side of his body. Some control of his muscles was gradually restored in the subsequent months, but he was left with grave handicaps. He could not walk without assistance, his left arm was not usable, and his speech was affected. His mind was not impaired, but it was the end of active scientific work. He might have been prepared to accept that as inevitable, since he had already (at the age of 86) slowed down considerably, but the restrictions on his movements were harder to bear. He had always been active and got his greatest satisfaction from doing things, and having now to spend most of each day confined to a chair was a terrible blow. He never gave up hope of being able to travel again and to attend international meetings, but one or two trips to London made by car with friends was the most that he was able to manage.

It was particularly disappointing to G.I. that he was not able to attend the 13th International Congress for Theoretical and Applied Mechanics held in Moscow in September 1972. These four-yearly Congresses are grand get-togethers for people working on the mechanics of fluids and solids, and G.I. always enjoyed them. He had attended and presented a paper at every one of the previous Congresses, beginning with the first in Delft in 1924. He had ready for presentation at the Congress in Moscow a piece of research on the making of holes in a thin sheet of liquid, and in the event it was read for him by Hugh Michael and incorporated in a later publication

Figure 17.1 *Near the end of active work.*

under their joint names (1973a). As noted in the previous chapter, this was a characteristically original contribution from G.I., involving simple mathematical arguments for the description of the effects of surface tension and some well-designed apparatus for the observation of equilibrium forms of a hole in a horizontal sheet of liquid. It was G.I.'s last piece of research, and was worthy of a place in his second golden period.

G.I. himself was missed by large numbers of participants at the Moscow Congress, and for many his absence was the ending of an era in mechanics. This widespread high regard for G.I. was a mixture of great admiration for his scientific accomplishments and affection for the man. He was universally liked, at home and abroad, and I do not know of a single person with whom he had an unfriendly relationship.

The period of three years following the stroke was one of frustration for G.I. He wrote letters, looked briefly at new issues of journals and was visited by friends, but it was difficult to fill in the time. Occasionally he was taken to have lunch or dinner at Trinity College, and although steps and stairs made such occasions hard work for him he welcomed them. He purchased an electrically driven wheelchair which could be controlled by one hand, and this enabled him to move round his garden by himself, but it could not negotiate gutters and hopes of going into Cambridge alone were not realized. During this time G.I. was cared for by Miss Gladys Davies, who had been with the Taylors as housekeeper since 1939. Miss Davies had provided the help needed by Stephanie during the years before her death in 1967, and now it was her devoted and skilled help which made life possible for G.I. It was sad for her and for the friends who visited him to see a noble spirit struggling against the inevitable deterioration and to realize that the one thing that would lift his spirits – involvement in scientific work of some kind – was now impossible.

G.I. was very conscious of his great debt to Gladys Davies and left 'Farmfield' to her in his will in order to provide for her in her old age. Gladys Davies died in July 1985, ten years after G.I., and she directed in her will that 'Farmfield' be sold and the proceeds be left to the University of Cambridge to establish a fund called the G.I. Taylor Memorial Fund. The University subsequently raised further money by appealing to educational and industrial institutions and individuals with an interest in fluid mechanics, the response of Trinity College where Taylor had been a Fellow for over 60 years being especially generous, and the two sums together enabled the University to establish a permanent professorship, the G.I. Taylor Professorship of Fluid Mechanics, as a lasting memorial to a great Cambridge scientist. The first holder of the G.I. Taylor Professorship was G.I. Barenblatt, who took up office on 1 October 1992. The second and current holder is T.J. Pedley.

A second although less severe stroke came in April 1975, and from then on G.I. was not often out of his bed. He was declining, and died two months later, nine months before his ninetieth birthday.

Apart perhaps from during these last three years, Geoffrey Taylor was a truly happy and contented man. He had spent a long life doing what he most wanted to do and doing it supremely well.

CHAPTER 18

The scientific legacy of G.I. Taylor

T he scientific legacy left to us by Geoffrey Taylor is two-fold in nature: it consists firstly of the superbly original and beautiful conceptions in his research into the mechanics of fluids and solids, and secondly of the brilliant demonstration of what a human mind can achieve.

With regard to the first, his published research contributions are reproduced in the four volumes of *The Scientific Papers of Sir Geoffrey Ingram Taylor*, and his papers and articles of all kinds are listed in chronological order at the end of this book. Many of these research developments have been described here, albeit without mathematical or experimental detail, and I hope they give as much pleasure to readers as they do to me. This first part of his legacy will stand for many years, and will suggest new further developments not yet conceived. It was characteristic of his papers that they opened up new areas of research for others to develop, thereby making this part of his legacy tangible.

The second and less evident part of Taylor's scientific legacy lies in the answer to the question, how did he do it? Are there some useful lessons for us in his approach to research and in his methods of working? He was outstandingly succcessful, and no doubt this was largely a consequence of his innate ability. Was it also due in part to human characteristics which the rest of us could usefully understand and adopt for the benefit of our own research? In order to be able to answer these difficult and subtle questions we need to consider what was characteristic of his approach and methods. With some hesitation I suggest that his research life had the following distinctive features, from which we might be able to extract some useful guidance.

(a) Perhaps the most striking characteristic of Taylor's research throughout his life was his passion for finding agreement between theoretical results and observations. He described this source of satisfaction with the restraint that was typical of him in his old age in the following words (1970a): 'The late Professor G.H. Hardy regarded all applied mathematics as a dull activity, a sort of glorified plumbing, which would not give the kind of satisfaction he found in pure mathematics. My feeling is that I derive a rather similar kind of satisfaction from the interplay between applied mathematics and experiment. It is quite a different kind from the satisfaction one gets in doing something useful, though one derives an added pleasure when anything one does turns out accidentally to be of use in engineering'.

It is easy to understand the intense pleasure given by the beautiful harmony of these two totally different modes of investigation, theory and observation, each one complementing the other and the two together making a near certainty that the theory is sound. Imagine, for instance, the excitement that Taylor presumably felt when he was observing the first appearance of cellular toroidal motion in the steady flow between two concentric cylinders. He had found, from a long calculation which rested on the assumption that eigen-modes of a disturbance exist – an assumption which was not universally accepted at that time – that, if the outer cylinder was held fixed and the speed of rotation of the inner cylinder was slowly raised, it would eventually reach a critical value above which the amplitude of a small disturbance sinusoidal in the axial direction would grow exponentially; and there was a value of the disturbance wave-number at which the growth rate was a maximum (1923c). He designed the apparatus with care, to ensure that the effect of the two ends of the annular region was small and the speed of rotation of the inner cylinder was accurately steady. In order to make the disturbance motion visible, he smeared the inner cylinder with dyed liquid (a typical clever idea). Then, when the speed of rotation of the inner cylinder slightly exceeded the critical value, there was the dramatic appearance of sheets of dye normal to the axis, regularly spaced in the axial direction and located at places where the disturbance velocity is normal to the axis and away from it. Two quantities had been both calculated and observed, one the critical rotation speed and the other the wave-number of the disturbance which becomes visible, and the agreement was good, for a whole range of values of the ratio of the two angular velocities of the two cylinders and of the two cylinder radii. I wish I had been there at such a moment!

This graphic demonstration that the stability of at least some steady flow systems could be calculated by normal-mode analysis had a major influence on subsequent work on hydrodynamic stability, but for Taylor the satisfaction lay in those sheets of dye appearing spontaneously at the 'right' rotation speed and with the 'right' spacing.

(b) Taylor always preferred to think in terms of concrete situations and specific problems. Unlike most other scientists with a mathematical training he saw less value in generalizations and abstract thinking. A good example of the way in which he illuminated a physical concept, namely, the expansion viscosity of a compressible fluid, was described in chapter 16. As a contribution to a discussion about the meaning and values of the two coefficients of viscosity, he pointed out the mechanism of dissipation when a medium consisting of small spherical gas bubbles dispersed in a liquid is being expanded, and showed how to calculate the expansion viscosity for such a medium (1954d). This explicit example of a medium for which the expansion viscosity is non-zero is a valuable aid to clear thinking about other such media.

Taylor described his preference for concrete problems in 'An unfinished dialogue with G.I. Taylor',[1] in response to the following question (which Taylor does not answer directly):

> **G.K.B.** Suppose that a young man who had just got his PhD came to you, in 1971, and said that he would like to choose a fruitful area of fluid mechanics and study it for a period of several years at least, and that he would like your advice on the best field to choose. Would you be able to offer him any guidance?
> **G.I.T.** I do not remember making any forecasts of broad areas of study which have proved fruitful, but I have gone along paths which are attractive to me personally. All my work, like that of most of us, has been concerned with particular problems. Some of these may point the way to a new range of particular problems, but I do not see how one can plan a 'strategy of research in fluid mechanics' otherwise than by thinking of particular problems. As you say, one may be directed to a

1. G.K. Batchelor, *J. Fluid Mech.* **70**, 1975, pp. 625–38.

particular field by social and political considerations but it seems to me that it is by attention to specific problems rather than by generalized reasoning that advances are made in our subject. I realize that by developing methods of analysis which have more general application than to the particular problems which give rise to them one may facilitate the solution of further problems, but *in general it seems to me it is through particular problems which can be subjected to experimental verification or compared with natural phenomena that most advances are made.*

The clause in italics (which are mine) puts Taylor's views in clear and definite form, and expresses an opinion on research strategy which I have not seen stated elsewhere and which seems to me to be worthy of study. There are obvious questions for consideration concerning the appropriateness of different research strategies for different fields of physical science. This is not the place for such a study; all we need note is the outstanding effectiveness of the approach adopted by Taylor throughout a long research life. 'Particular' and 'special' are regarded as slightly pejorative adjectives by most applied mathematicians, but Taylor's matchless insight came from a consideration of the physical processes at work in actual particular cases.

In a memoir on von Kármán (1973b), G.I. ascribed the same view to him: 'We both of us preferred special problems and models to the more general themes which some mathematicians, after writing down the relevant equations, seem to leave hanging in the air with the implication that any competent engineer can do the rest.'

(c) Taylor's astonishing independence and self-containment should be included in this list of characteristics which bear on his work. He was not a product of his time or of his place, and, as was said of Rutherford, he could have done physics at the North Pole had he wished. The first striking demonstration of this independence in his work was the decision *not* to go further with research on 'modern physics' in 1908 after completing his vacation experiment on interference fringes with feeble light. It was a time of dramatic discoveries and novel theories in atomic physics in particular, and one would have supposed that the prospect of working under J. J. Thomson or one of his colleagues in the Cavendish Laboratory in 1908 would have had a powerful appeal for a young research student. As to his thinking at this time we know only that many years later he wrote (1963c) 'I did not feel a call to a career in pure physics'. Instead he turned his attention to the structure of shock waves, a choice which possibly was made after seeing a paper on that topic by Rayleigh published in 1908.

Then there was the bold decision as an inexperienced meteorologist of age 27 to spend the spring and summer months of 1913 serving on the *Scotia* expedition and making measurements of velocity, temperature and water vapour content in the atmosphere up to great heights. Many instances of a tendency to think things through for himself and to make his own decisions without being unduly influenced by the views of others could be given. His independence of spirit found fulfilment in his passion for sailing in small boats, this being an activity that combined adventure with control of the elements – wind and waves – for which he had a close affinity. The invention of a better anchor than one whose traditional design had been accepted for centuries undoubtedly required him first to clear his mind of conventional ideas. A practical problem needed solution and the independent thought that this required came naturally to him.

(d) The fact that Taylor was a happy man shines through his work. He would have known what Noel Coward meant when he said 'Work is much more fun than fun.' He was friendly and unassuming and likeable, and had an uncomplicated character. I think it would be fair to describe him as emotionally simple and unreflective. He had a razor sharp mind but it was not much used on political issues, social affairs, teaching or serious literature (to judge by the books left in his house when he died). He kept his sharp wits for the one thing that interested him above all others, namely, the understanding of mechanical phenomena. It was no doubt a consequence of his simple character and lack of involvement in complex issues that he was able to approach his research so naturally and so happily, free from maladjustments and self-concern. Such men do not normally make interesting companions, and he was certainly not a good conversationalist, but those who managed to get behind the gentle shy manner, especially on paper where Taylor's intelligence and dry humour came into their own, found an entirely lovable man with a zest for life which for me is typified by the letter reproduced as figure 18.1.

(e) The word 'simplicity' becomes over-worked in a description of the characteristic features of Taylor's research – and indeed of his life. This ability to see how to do things simply and economically and clearly was especially evident in his experimental work. He had a gift for designing an experiment in such a way that it yielded the required information with a minimum of time, cost, and error, and a maximum of clarity. He seemed to be able to see immediately how to make the experimental conditions correspond with those assumed in the corresponding theory. An example of the way in which Taylor's originality enabled him to achieve this simplicity may be illuminating.

In 1954 Taylor was thinking about mass transport in tubes (1954e), and among other theoretical results he calculated the critical vertical gradient of density of salt solution which remains at rest in a vertical tube. For density gradients below this critical value the fluid is stable and remains stationary owing to molecular transport effects; above it the fluid is unstable and convective over-turning sets in. As always Taylor wanted to confirm experimentally his calculation of the critical density gradient. How should it be done? If one took one's cue from the numerous experiments with Bénard cells, one would think of beginning with a small density gradient for which the system is stable and then gradually increasing the gradient by some process involving diffusion until overturning occurs. But this is hopelessly slow, and moreover it would be difficult to generate a density distribution which has a uniform gradient over an appreciable vertical distance. Taylor saw, and I do not think any period of conscious thought was required, that the 'right' way to do the experiment is to *begin* with a supercritical gradient and allow the resulting convection to diminish the gradient until it is no longer supercritical. One simply connects the top of a vertical tube which initially is full of pure water to a reservoir of dyed salt solution of known (high) concentration, and then one waits until the dye has stopped being carried down the tube by the over-turning that occurs under conditions of instability. The observation now consists of nothing more than measuring the length of the column of dye with a ruler! One feels one could have thought of such a beautifully

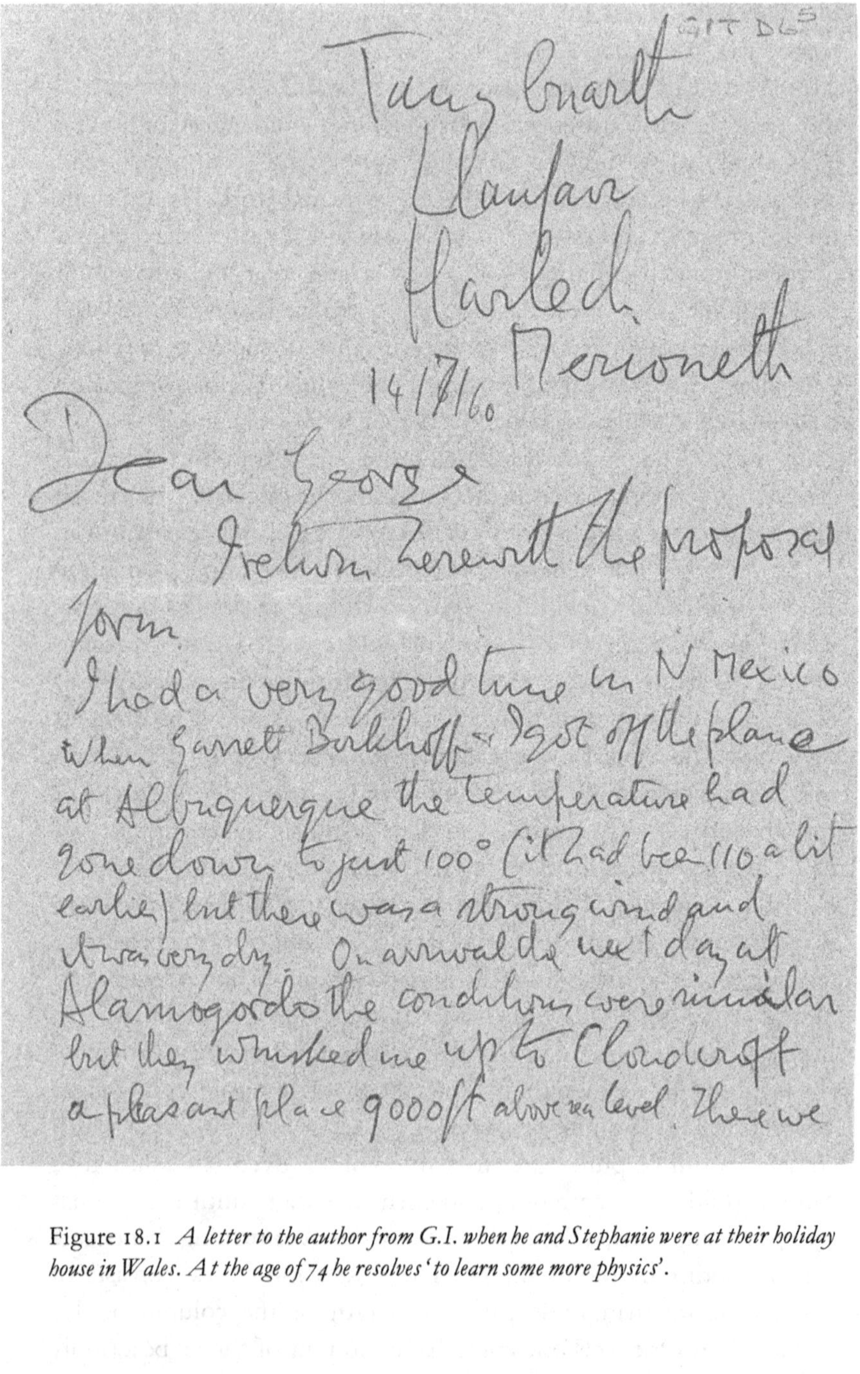

Tan y Craig
Llanfair
Harlech, Merioneth
14/7/60

Dear George
    I reclose herewith the proposal
form.
    I had a very good time in N Mexico
When Garrett Birkhoff + I got off the plane
at Albuquerque the temperature had
gone down to just 100° (it had been 110 a bit
earlier) but there was a strong wind and
it was very dry.   On arrival the next day at
Alamogordo the conditions were similar
but they whisked me up to Cloudcroft
a pleasant place 9000ft above sea level. There we

Figure 18.1 *A letter to the author from G.I. when he and Stephanie were at their holiday house in Wales. At the age of 74 he resolves 'to learn some more physics'.*

had our symposium which was really a summer school attended by people from all over U.S.A. & I had 3 really first rate colleagues HP Robertson, Joseph Bardeen & Fowler. Robertson gave a course on relativity & Fowler (so called) on nuclear physics & its connection with the history of the universe. These two sets of lectures were really fascinating & made me resolve to learn some more physics. Bardeen gave a set of lectures on superconductivity which were I think pretty conclusive though I do not know enough to be able to understand most of them — and he is not nearly such a good lecturer as the other two.

Anyhow a good time was had by all and I came back with enough dollars to pay for my car.

Jim & I. 7.

simple idea; but Taylor's originality usually enabled him to be the first to do so.

(f) I have not yet mentioned insight explicitly. How was it that he seemed able to see instantly what was going on? Harry Jones had this to say: 'His intuition seemed to be based not so much on a great accumulation of empirical knowledge as on a deep understanding of the laws of physics.'

(g) Finally, there is a feature of Taylor's work which I can only describe as 'having significance'. By this I mean that he seemed always to be able to recognize and understand the essential aspects of a phenomenon or a problem that everyone sees later to be funda-mental and of wide applicability. Some observations are significant, some are not; and the same may be said of theoretical results. Similarity solutions are usually significant; and no-one was quicker to spot the existence of a similarity solution than Taylor (see 1960c for some examples). Asymptotic results likewise contain much illu-minating information, and Taylor exploited them to the full. Sir Arnold Hall, one of Taylor's research students in the late thirties, recognized the significance of much of Taylor's work, and described it as follows: 'He had a great sense of the physical situation. His immediately available mathematical equipment, although formid-able, was perhaps less developed than that of some others like von Kármán, but he could turn to the textbooks and rapidly master any mathematical area necessary for his work. His intuitive feeling for experiment was strong, and he was able to devise a route for experimental investigation with great elegance, and often with great rapidity. This was during a period – which has gone on ever since – in which those of a mathematical bent could set themselves almost an infinity of problems and then proceed to solve them. Taylor was not like that – he had an intense sense of what really mat-tered, and of what was at the heart of practical and serious problems in science and engineering. It was for this reason that his work was of immensely greater long-term significance than that of most of his contemporaries'.

One could no doubt measure the total 'significance' of Taylor's work in a rough way by listing some of the developments to which his name has become attached by the usual mysterious process:

Taylor–Couette instability (*chapter 7*)
Taylor–Proudman theorem (*chapter 7*)
Taylor column (*chapter 7*)
Taylor dislocations (*chapter 11*)
Taylor vorticity transfer theory (*chapter 12*)
Taylor–Green problem (*chapter 12*)
Taylor microscale (*chapter 12*)
Taylor frozen-flow hypothesis (*chapter 12*)
Rayleigh–Taylor instability (*chapter 15*)
Taylor dispersion (*chapter 16*)
Saffman–Taylor fingering (*chapter 16*)

Originality, independence, insight, and the ability to recognize significance, are of course related and not easily separated. Taylor had them all in outstanding measure and used them to the full. And there was naturalness. I have never known a man to fit so naturally into the pattern of his life. Taylor's demonstration of the value of naturalness in his life and in his work is what I appreciate most in this second part of his legacy.

An applied mathematician's apology[1]
(1956f)

When our President wrote saying that you propose to honour me by giving me the de Morgan Medal I was of course very pleased and I must confess not a little surprised because my contributions to study of what G.H. Hardy called 'real mathematics' have been small. I realise that you recognise in me a person who, however elementary his technique, has always liked to think of his problems in a mathematical way. I am particularly glad to have my name connected in this way with Augustus de Morgan for he was a friend and correspondent of my grandfather George Boole. It seems to me that de Morgan had a place in mathematics rather like that of Huxley in biology. Huxley was a forceful expounder of views which were then unacceptable to some of the bishops who attacked him in print and to whom he made vigorous replies. De Morgan upheld the advantages of mathematical ways of thinking, particularly about logic, and was violently attacked by a Scottish philosopher Sir William Hamilton (not of course Rowan Hamilton) who poured ridicule on the idea that mathematics has anything to tell the philosophers.

In more recent times the study of mathematics has been regarded by the non-mathematical world as part of science rather than philosophy and it has been attacked as leading to a knowledge of nature which is undesirable because it can be misused by evil men. It was, I think, a feeling that the outside world has no idea of why it is that a mathematician does what he does that inspired Hardy to write his delightful *Mathematician's Apology*, a book which I expect most people here have read. In it Hardy describes his own life as a mathematician and

1. In 1956 G.I. was awarded the de Morgan medal of the London Mathematical Society. This is his address in reply, with an ending which may be incomplete.

tells of the conversion which he experienced on reading Jordan's Cours d'Analyse from being a clever problem solver to becoming a 'real mathematician' and he explains what he means by that phrase. He distinguishes between the mental activity of a chess player who solves mathematical problems of great complexity using only intricate combinations of a few simple rules and that of a real mathematician who is driven to delve ever more deeply into the foundations of his subject. For a piece of analysis to be satisfying to a real mathematician it must, in Hardy's view, have depth as well as beauty and he tried to give an idea of what he meant by both these concepts. His apology for his life as a mathematician was roughly that he was pursuing a purely aesthetic aim and that in this pursuit neither he nor his fellow mathematicians ever did any harm to anyone.

In describing the aesthetic quality of his pleasure in mathematics Hardy had to express aesthetic preferences and in matters of taste there is no basis for dispute. He had, I believe, little appreciation of any kind of beauty outside the field of pure mathematics, and that may partly explain his dislike and perhaps misunderstanding of applied mathematics. His view seemed to be that the question whether a mathematical theorem has any bearing on the physical world is entirely irrelevant so far as its aesthetic content is concerned. He would, I think, have classed together such problems as those of the stability of fluid flow or the mathematical representation of large strains with problems about weightless strings attached to perfectly rough elephants on perfectly smooth planes. He classed such things indiscriminately as school mathematics which he regarded as hopelessly dull.

If I were to attempt an apology for my own life as an applied mathematician I would follow Hardy's lead in saying that the satisfaction which we have in our work – on the rare occasions when it comes off – is essentially aesthetic. It is the pleasure we experience in finding out how different observable phenomena of the physical world fit together. As Hardy points out most of the mathematics we use is what he calls school mathematics. To people less gifted than Hardy even that may be interesting, but to us it is the discovering of the connection between physical phenomena and describing them by mathematical analysis, rather than the analysis itself, which is interesting. I thought therefore that in addressing an audience mainly of pure mathematicians it might be appropriate to describe some of the simple or even trivial things that have pleased or amused me in my time.

First let me remind you of one of the most beautiful and at the same time one of the most simple ideas that ever came to an applied mathematician, the principle of Archimedes. This can be expressed, as is sometimes done in text books of hydrostatics, as a relationship involving surface and volume integrals. In that form it does seem rather dull, but as conceived by Archimedes in his bath it has exactly the quality which simple-minded folk like me find so satisfying. I remember once, when I had to set an examination question, thinking up the following straightforward example of its application: if a solid sphere – say a wooden ball – floats on the surface of a fluid which is rotating like a solid body about a vertical axis, it will gravitate towards the lowest point, which is on the axis of rotation. If however the sphere be weighted by fixing a piece of lead to its surface till it nearly sinks the sphere will travel outward and therefore upwards till it strikes the wall of the vessel containing the fluid. Anyone who really understands Archimedes' principle will see at once that this is true, but I must confess that one of my fellow examiners was unconvinced till I took him into my laboratory and showed him weighted and unweighted wooden balls floating on water in a rotating basin and doing just what was predicted.

It would be difficult to convey in a short lecture any idea of the methods and aims of applied mathematicians; they are too varied. All I can do is describe by way of illustration the sequence of thoughts which has guided one line of my own investigations, which is still under way. I fear it may only show you that I am not a 'real mathematician' in Hardy's sense, but I do hope to give an idea of the kind of satisfaction we get out of the continual checking of our mathematical results by experiment.

About two years ago I had a letter from a young man working in the research department of a big paper company in Canada asking for my help in thinking about the way in which water is drained from paper pulp in a paper-making machine. Paper was originally made by shaking a pulp of fibres and water in a tray with a porous bottom. The water drained away through the action of gravity but since the head of water was very small the drainage was a slow process. During the French revolution it was found that democracy will not work without much more paper than could be made in that way, so the continuous process was devised. The pulp is projected through a narrow horizontal slit on to a moving band made of wire gauze. The pulp drains as the band moves and when it reaches a consistency which gives it strength enough, it is peeled off the band. One would have thought that the time taken for the requisite amount of drainage would not depend on the speed of the band, so that its length would have to be increased proportionally to the speed of working if there is to be enough drainage to solidify the paper before reaching the place where it is peeled off the band. Experience however showed that this is not the case. The reason was found to be that the band was supported on a series of rollers designed merely to keep it flat, and the hydrodynamic flow in the narrow space between the roller and the band near their line of contact gives rise to a large suction which is enormously more effective than gravity in draining the pulp. Though millions have been spent on building paper machines no one seems to have reasoned effectively about the actual mechanism of their operation till a few years ago. Recently several theories have been proposed and I was asked to give an opinion as to the validity of some of them. It seemed to me that one of them was essentially right, but on examining it I found that it rested on two necessary, but mutually inconsistent, assumptions.

G.I. Taylor: honours

A complete record of the honours received by Taylor was not kept, and the following list may be imperfect.

(i) *Honorary degrees*

    1930 University of Aachen
    1933 University of Liverpool
    1933 University of British Columbia
    1938 University of Oxford
    1946 University of London
    1946 University of Oslo
    1952 University of Birmingham
    1952 University of Istanbul
    1953 University of Edinburgh
    1957 University of Cambridge
    1959 University of Bristol
    1961 University of Paris
    1964 Technical University of Milan
    1967 University of Michigan
    1971 Colorado State University

(ii) *Election to membership or fellowship of learned societies*

    1919 Royal Society
    1938 Royal Society of Edinburgh
    1939 Institute of Aeronautical Sciences of America
    1940 Royal Netherlands Academy of Sciences
    1945 Institution of Mechanical Engineers
    1945 US National Academy of Sciences
    1946 Academy of Sciences of Norway
    1946 Académie des Sciences, Paris

1947 Institute of Metals
1948 Royal Aeronautical Society
1948 Calcutta Mathematical Society
1951 Accademia dei Lincei, Rome
1952 Indian Academy of Sciences
1955 Institution of Civil Engineers
1955 Swedish Royal Academy of Sciences
1955 American Philosophical Society
1956 American Academy of Arts and Sciences
1959 Manchester College of Science and Technology
1960 International Academy of Aeronautics
1961 Turin Academy of Science
1962 Franklin Institute
1962 American Meteorological Society
1963 American Institute of Aeronautics and Astronautics
1964 Institute of Physics and Physical Society
1965 Institute of Mathematics and its Applications
1966 Academy of Sciences of USSR
1966 American Society of Mechanical Engineers
1970 Polish Academy of Sciences

(iii) *Awards and medals*

1915 Adams Prize, University of Cambridge
1918 Hawksley Medal, Institution of Mechanical Engineers
1933 Royal Medal, Royal Society
1935 King's Silver Jubilee Medal
1940 Ewing Medal, Institution of Civil Engineers
1944 Knighthood
1944 Copley Medal, Royal Society
1946 US Medal for Merit
1951 Symons Medal, Royal Meteorological Society
1953 Royal Coronation Medal
1954 Gold Medal, Royal Aeronautical Society
1954 Exner Medal, Oesterreichischer Gewerbeverein
1956 de Morgan Medal, London Mathematical Society

1958   Panetti Prize and Medal, Accademia della Scienze di Torino

1958   Dutch Medal, Akkademia van Wetenschafen

1958   Timoshenko Medal, American Society of Mechanical Engineers

1959   Kelvin Medal, Institution of Civil Engineers

1962   Trasenster Medal, Université de Liege

1962   Franklin Medal, Franklin Institute

1962   Albert Sauveur Achievement Award, American Society of Metals

1962   Faraday Award

1962   Medal of Honor, Rice University

1964   Platinum Medal, Institute of Metals

1965   James Watt Medal, Institution of Mechanical Engineers

1969   Order of Merit

1969   von Kármán Medal, American Society of Civil Engineers

1969   A.A. Griffith Medal, Brunel University

1970   Jubilee Medal, American Meteorological Society

1972   von Kármán Prize, Society of Industrial and Applied Mathematics

1973   Medal of Honour, Danish Metallurgical Society

Articles about G.I. Taylor

Anonymous

1945 Sir Geoffrey Taylor. *Monthly Science News.*

1957 Sir Geoffrey Taylor. His life's rule: to do what interests him. *New Scientist.*

1975 Sir Geoffrey Taylor. Obituary, *The Times.*

G.K. Batchelor

1964 Close-up: Sir Geoffrey Taylor. *Trinity Review.*

1975 An unfinished dialogue with G.I. Taylor. *J. Fluid Mech.* **70**, 625–638.

1976 G.I. Taylor as I knew him. *Advances in Appl. Mech.* **16**, 1–8.

1976 Geoffrey Ingram Taylor, 1886–1975. *Biog. Mem. Fell. Roy. Soc.* **22**, 565–633.

1986 Geoffrey Ingram Taylor, 7 March 1886 – 27 June 1975. *J. Fluid Mech.* **178**, 1–14.

Educational Services Inc.

1967 An Interview with G.I. Taylor. 16 mm colour & sound.

T. Griffiths

1972 G.I. Taylor – a profile. *Trinity Review.*

R.V. Southwell

1956 G.I. Taylor: a biographical note. *Surveys in Mechanics,* Cambridge University Press, 1–6.

D.B. Spalding

1962 An interview with Sir Geoffrey Taylor. *The Chartered Mechanical Engineer,* **9**, 186–191.

Bibliography of works by G.I. Taylor

Almost all Taylor's scientific papers, published and unpublished, have been reprinted in four volumes with the title *The Scientific Papers of Sir Geoffrey Ingram Taylor*, published by Cambridge University Press. The designation SP IV, 1 in the list below indicates that a paper has been reprinted in volume IV and is paper number 1 within that volume.

1909a	Interference fringes with feeble light. *Proc. Camb. Phil. Soc.* **15**, 114–15.	(SP IV, 1)
1910a	The conditions necessary for discontinuous motion in gases. *Proc. R. Soc. Lond.* A **84**, 371–7.	(SP III, 1)
1914a	Report on the work carried out by the s.s. *Scotia*, 1913, pp. 48–68. London: HMSO.	
1915a	Eddy motion in the atmosphere. *Phil. Trans. R. Soc. Lond.* A **215**, 1–26.	(SP II, 1)
b	The use of fin surface to stabilize a weight towed from an aeroplane. *Rep. Memo. Advis. Comm. Aeronaut.* no. 184.	(SP III, 2)
c	Report on the accuracy with which temperature errors in determining heights by barometer may be corrected. *Rep. Memo. Advis. Comm. Aeronaut.* no. 239.	
1916a	Skin friction of the wind on the earth's surface. *Proc. R. Soc. Lond.* A **92**, 196–9.	(SP II, 2)
b	Conditions at the surface of a hot body exposed to the wind. *Rep. Memo. Advis. Comm. Aeronaut.* no. 272.	(SP II, 3)
c	On the dissipation of sound in the atmosphere. Paper for Advis. Comm. Aeronaut.	(SP II, 4)
d	(With C. J. P. Cave) Variation of wind velocity close to the ground. *Rep. Memo. Advis. Comm. Aeronaut.* no. 296, part 1.	(SP II, 5)
e	Pressure distributio round a cylinder. *Rep. Memo. Advis. Comm. Aeronaut.* no. 191	(SP III, 3)
f	Pressure distribution over the wing of an aeroplane in flight. *Rep. Memo. Advis. Comm. Aeronaut.* no. 287.	(SP III, 4)
g	Phenomena connected with turbulence in the lower atmosphere. *Rep. Memo. Advis. Comm. Aeronaut.* no. 304.	

1917a   (With A.A. Griffith) The use of soap films in solving torsion problems. *Rep. Memo. Advis. Comm. Aeronaut.* no. 333, and *Proc. Inst. Mech. Eng.*, pp. 755–89.   (SP I, 1)

    b   (With A.A. Griffith). The problem of flexure and its solution by the soap-film method. *Rep. Memo. Advis. Comm. Aeronaut.* no. 399.   (SP I, 2)

    c   The formation of fog and mist. *Q. Jl R. Met. Soc.* **43**, 241–68.   (SP II, 6)

    d   Observations and speculations on the nature of turbulent motion. *Rep. Memo. Advis. Comm. Aeronaut.* no. 345.   (SP II, 7)

    e   Phenomena connected with turbulence in the lower atmosphere. *Proc. R. Soc. Lond.* A **94**, 137–55.   (SP II, 8)

    f   Motion of solids in fluids when the flow is not irrotational. *Proc. R. Soc. Lond.* A **93**, 99–113.   (SP IV, 2)

    g   Fog conditions. *Aeronaut. J.* **21**, 75–90.

1918a   (With A.A. Griffith) The application of soap films to the determination of the torsion and flexure of hollow shafts. *Rep. Memo. Advis. Comm. Aeronaut.* no. 392.   (SP I, 3)

    b   On the dissipation of eddies. *Rep. Memo. Advis. Comm. Aeronaut.* no. 598.   (SP II, 9)

    c   Skin friction on a flat surface. *Rep. Memo. Advis. Comm. Aeronaut.* no. 604.   (SP II, 10)

1919a   Tidal friction in the Irish Sea. *Phil. Trans. R. Soc. Lond.* A **220**, 1–33.   (SP II, 11)

    b   On the shapes of parachutes. Paper for Advis. Comm. Aeronaut.   (SP III, 5)

1920a   Tidal friction and the secular acceleration of the moon. *Mon. Not. R. Astr. Soc.* **80**, 308–9.   (SP II, 12)

    b   Navigation notes on a passage from Burnham-on-Crouch to Oban. *Yachting Monthly.*

1921a   Tidal oscillations in gulfs and rectangular basins. *Proc. Lond. Math. Soc.* **20**, 148–81.   (SP II, 13)

    b   Diffusion by continuous movements. *Proc. Lond. Math. Soc.* **20**, 196–212.   (SP II, 14)

    c   Tides in the Bristol Channel. *Proc. Camb. Phil. Soc.* **20**, 320–5.   (SP II, 15)

    d   Scientific methods in aeronautics. *Aeronaut. J.* **25**, 474–91.   (SP III, 6)

    e   The 'rotational inflow factor' in propeller theory. *Rep. Memo. Aeronaut. Res. Comm.*, no. 765.   (SP III, 7)

    f   Experiments with rotating fluids. *Proc. R. Soc. Lond.* A **100**, 114–21.   (SP IV, 3)

    g   Experiments with rotating fluids. *Proc. Camb. Phil. Soc.* **20**, 326–9.

1922a   A relation between Bertrand's and Kelvin's theorems on impulses. *Proc. Lond. Math. Soc.* **21**, 413–14.   (SP I, 4)

1922b   Notes on Mr. Glauert's paper, 'An aerodynamic theory of the air-screw'. Paper for Aeronaut. Res. Comm.   (SP III, 8)

1922c   The motion of a sphere in a rotating liquid. *Proc. R. Soc.*
      Lond. A **102**, 180–9.   (SP IV, 4)

1923a   (With C.F. Elam) The distortion of an aluminium crystal
      during a tensile test. *Proc. R. Soc. Lond.* A **102**, 643–67.   (SP I, 5)

    b   The decay of eddies in a fluid. Paper for Aeronaut. Res.
      Comm.   (SP II, 16)

    c   Stability of a viscous liquid contained between two rotating
      cylinders. *Phil. Trans. R. Soc. Lond.* A **223**, 289–343.   (SP IV, 5)

    d   The motion of ellipsoidal particles in a viscous fluid. *Proc. R.*
      Soc. Lond. A **103**, 58–61.   (SP IV, 6)

    e   On the decay of vortices in a viscous fluid. *Phil. Mag.* **46**,
      671–4.   (SP IV, 7)

    f   Experiments on the motion of solid bodies in rotating
      fluids. *Proc. R. Soc. Lond.* A **104**, 213–18.   (SP IV, 8)

1924a   The singing of wires in a wind. *Nature, Lond.* **113**, 536.   (SP III, 9)

    b   Extracts from the log of *Frolic. R. Cruising Club. F.*, pp. 85–
      105.

1925a   (With W.S. Farren) The heat developed during plastic
      extension of metals. *Proc. R. Soc. Lond.* A **107**, 422–51.   (SP I, 6)

    b · (With C.F. Elam) The plastic extension and fracture of alu-
      minium crystals. *Proc. R. Soc. Lond.* A **108**, 28–51.   (SP I, 7)

    c   Notes on the 'Navier effect'. Paper for Aeronaut. Res.
      Comm.   (SP I, 8)

    d   Note on the connection between the lift on an aerofoil in a
      wind and the circulation around it. *Phil. Trans. R. Soc. Lond.*
      A **225**, 238–45.   (SP III, 10)

    e   (With C.T.R. Wilson) The bursting of soap-bubbles in a
      uniform electric field. *Proc. Camb. Phil. Soc.* **22**, 728–30.   (SP IV, 9)

    f   Experiments with rotating fluids. *Proc. 1st Int. Congr. Appl.*
      Mech., Delft, 1924, pp. 89–96.

    g   Versuche mit rotierenden Flüssigkeiten. *Z. Angew. Math.*
      Mech. **5**, 250–3.

1926a   (With W.S. Farren) The distortion of crystals of aluminium
      under compression. I. *Proc. R. Soc. Lond.* A **111**, 529–51.   (SP I, 9)

    b   (With C.F. Elam) The distortion of iron crystals. *Proc. R.*
      Soc. Lond. A **112**, 337–61.   (SP I, 10)

1927a   The distortion of single crystals of metals. *Proc. 2nd Int.*
      Congr. Appl. Mech., Zürich, 1926, pp. 46–52.   (SP I, 11)

    b   The distortion of crystals of aluminium under compression.
      II. Distortion by double slipping and changes in orientation
      of crystal axes during compression. *Proc. R. Soc. Lond.* A **116**,
      16–38   (SP I, 12)

    c   The distortion of crystals of aluminium under compression.
      III. Measuremnents of stress. *Proc. R. Soc. Lond.* A **116**, 39–
      60.   (SP I, 13)

1927d   An experiment on the stability of superposed steams of
fluid. *Proc. Camb. Phil. Soc.* **23**, 730–1.   (SP II, 17)

     e   Turbulence. *Q. J. R. Met. Soc.* **53**, 201–11.

     f   Across the Arctic circle in *Frolic*, 1927. R. *Cruising Club J.*,
pp. 9–26.

1928a   The deformation of crystals of β-brass. *Proc. R. Soc. Lond.* A
118, 1–24.   (SP I, 14)

     b   Resistance to shear in metal crystals. *Trans. Faraday Soc.* **24**,
121–5.   (SP I, 15)

     c   A manometer for use with small Pitot tubes. *Proc. Camb.
Phil. Soc.* **24**, 74–5.   (SP III, 11)

1928d   The energy of a body moving in an infinite fluid, with an
application to airships. *Proc. R. Soc. Lond.* A **120**, 13–21.   (SP III, 12)

     e   The forces on a body placed in a curved or converging
stream of fluid. *Proc. R. Soc. Lond.* A **120**, 260–83.   (SP III, 13)

     f   (With C.F. Sharman) A mechanical method for solving
problems of flow in compressible fluids. *Proc. R. Soc. Lond.* A
121, 194–217.   (SP III, 14)

     g   The force acting on a body placed in a curved and conver-
ging stream of fluid. *Rep. Memo. Aeronaut. Res. Comm.* no.
1166.

     h   Report on progress during 1927–8 in calculation of flow of
compressible fluids and suggestions for further work. *Rep.
Memo. Aeronaut Res. Comm.* no. 1196.

1929a   The criterion for turbulence in curved pipes. *Proc. R. Soc.
Lond.* A **124**, 243–9.   (SP II, 18)

     b   Waves and tides in the atmosphere. *Proc. R. Soc. Lond.* A **126**,
169–83.   (SP II, 19)

     c   The air wave from the great explosion at Krakatau. *4th
Pacific Science Congr., Java*, 1929, vol. II B, 645–55.

     d   Visit to Japan. 16 mm film, silent black and white.

1930a   The application of Osborne Reynold's theory of heat trans-
fer to flow through a pipe. *Proc. R. Soc. Lond.* A **129**, 25–30.   (SP II, 20)

     b   The flow of air at high speeds past curved surfaces. *Rep.
Memo. Aeronaut. Res. Comm.* no. 1381.   (SP III, 15)

     c   Some cases of flow of compressible fluids. *Rep. Memo.
Aeronaut. Res. Comm.* no. 1382.   (SP III, 16)

     d   Recent work on the flow of compressible fluids. *J. Lond.
Math. Soc.* **5**, 224–40.   (SP III, 17)

     e   Tour in the East Indies. *Proc. R. Instn* **26**, 209.

1930f   Stromung um einen Korper in einer kompressiblen
Flussigkeit. *Z. Angew. Math. Mech.* **10**, 334–45.

1931a   (With H. Quinney) The plastic distortion of metals. *Phil.
Trans. R. Soc. Lond.* A **230**, 323–62.   (SP I, 16)

1931b Effect of variation in density on the stability of superposed
streams of fluid. *Proc. R. Soc. Lond.* A **132**, 499–523.   (SP II, 21)

   c Internal waves and turbulence in a fluid of variable density.
Rapp. P.-V. Réun. Cons. Int. Explor. Mer 76, 35–42.   (SP II, 22)

   d The flow round a body moving in a compressible fluid. *Proc.
3rd Int. Cong. Appl. Mech., Stockholm, 1930*, vol. I, 263–275.

   e Round Ireland in *Frolic*. *R. Cruising Club J.*, pp. 213–225.

1932a (With H. Quinney) The distortion of wires on passing
through a draw-plate. *J. Inst. Metals* **49**, 187–99.   (SP I, 17)

   b Note on the distribution of turbulent velocities in a fluid
near a solid wall. *Proc. R. Soc. Lond.* A **135**, 678–84.   (SP II, 23)

   c The transport of vorticity and heat through fluids in turbulent motion. *Proc. R. Soc. Lond.* A **135**, 685–705.   (SP II, 24)

   d The resonance theory of semidiurnal atmospheric
oscillations. *Mem. R. Met. Soc.* **4**, 43–51.   (SP II, 25)

   e Applications to aeronautics of Ackeret's theory of supersonic aerofoils moving at speeds greater than that of sound.
Rep. Memo. Aeronaut. Res. Comm. no. 1467.   (SP III, 18)

   f The viscosity of a fluid containing small drops of another
fluid. *Proc. R. Soc. Lond.* A **138**, 41–8.   (SP IV, 10)

   g Note on review by Davies and Sutton of the present position of the theory of turbulence. *Q. J. R. Met. Soc.* **58**, 61–5.

1933a The buckling load for a rectangular plate with four clamped
edges. *Z. Angew. Math. Mech.* **13**, 147–52.   (SP I, 18)

   b (With J.W. Maccoll) The air pressure on a cone moving at
high speeds. I. *Proc. R. Soc. Lond.* A **139**, 278–97.   (SP III, 19)

   c (With J.W. Maccoll) The air pressure on a cone moving at
high speeds. II. *Proc. R. Soc. Lond.* A **139**, 298–311.   (SP III, 20)

   d (With J.W. Maccoll) L'onde ballistique d'un projectile à tête
conique. *Mém. de l'Art. Franç.* **12**, 651–83.

1934a (With H. Quinney) The latent energy remaining in a metal
after cold working. *Proc. R. Soc. Lond.* A **143**, 307–26.   (SP I, 19)

   b Faults in a material which yields to shear stress while
retaining its volume elasticity. *Proc. R. Soc. Lond.* A **145**, 1–
18.   (SP I, 20)

   c The mechanism of plastic deformation of crystals. I.
Theoretical. *Proc. R. Soc. Lond.* A **145**, 362–87.   (SP I, 21)

   d The mechanism of plastic deformation of crystals. II.
Comparison with observations. *Proc. R. Soc. Lond.* A **145**,
388–404.   (SP I, 22)

   e The strength of rock salt. *Proc. R. Soc. Lond.* A **145**, 405–15.   (SP I, 23)

   f A theory of the plasticity of crystals. *Z. Kristall.* A **89**, 375–
85.   (SP I, 24)

   g The formation of emulsions in definable fields of flow. *Proc.
R. Soc. Lond.* A **146**, 501–23.   (SP IV, 11)

1934h The holding power of anchors. *Yachting Monthly and Motor Boating Mag.* (SP IV, 12)

1935a Lattice distortion and latent heat of cold work in copper. Paper for Aeronaut. Res. Comm. (SP I, 25)

b Turbulence in a contracting stream. *Z. Angew. Math. Mech.* **15**, 91–96. (SP II, 26)

c Statistical theory of turbulence. I. *Proc. R. Soc. Lond.* A **151**, 421–44. (SP II, 27)

d Statistical theory of turbulence. II. *Proc. R. Soc. Lond.* A **151**, 444–54. (SP II, 28)

e Statistical theory of turbulence. III. Distribution of dissipation of energy in a pipe over its cross-section. *Proc. R. Soc. Lond.* A **151**, 455–64. (SP II, 29)

f Statistical theory of turbulence. IV. Diffusion in a turbulent air stream. *Proc. R. Soc. Lond.* A **151**, 465–78. (SP II, 30)

g Distribution of velocity and temperature between concentric rotating cylinders. *Proc. R. Soc. Lond.* A **151**, 494–512. (SP II, 31)

h (With J.W. Maccoll) The mechanics of compressible fluid. Section H of *Aerodynamic theory*, vol. III (ed. W.F. Durand), pp. 209–50. Berlin: Springer.

i Orbituary of Sir Horace Lamb. *Nature* 16 Feb. 1935, 255–7.

1936a The mean value of the fluctuations in pressure and pressure gradient in a turbulent fluid. *Proc. Camb. Phil. Soc.* **32**, 380–4. (SP II, 32)

b Statistical theory of turbulence. V. Effect of turbulence on boundary layer. Theoretical discussion of relationship between scale of turbulence and critical resistance of spheres. *Proc. R. Soc. Lond.* A **156**, 307–17. (SP II, 33)

c The oscillations of the atmosphere. *Proc. R. Soc. Lond.* A **156**, 318–26. (SP II, 34)

d Correlation measurements in a turbulent flow through a pipe. *Proc. R. Soc. Lond.* A **157**, 537–46. (SP II, 35)

e Fluid friction between rotating cylinders. I. Torque measurements. *Proc. R. Soc. Lond.* A **157**, 546–64. (SP II, 36)

f Fluid friction between rotating cylinders. II. Distribution of velocity between concentric cylinders when outer one is rotating and inner one is at rest. *Proc. R. Soc. Lond.* A **157**, 565–78. (SP II, 37)

g Well established problems in high speed flow. *R. Accad. Ital. Att.: 5, Conv. Sci. Fis. Mat. Nat.*, pp. 198–214.

1937a (With H. Quinney) The emission of the latent energy due to previous cold working when a metal is heated. *Proc. R. Soc. Lond.* A **163**, 157–81. (SP I, 26)

b (With A.E. Green) Mechanism of the production of small eddies from large ones. *Proc. R. Soc. Lond.* A **158**, 499–521. (SP II, 38)

1937c Flow in pipes and between parallel planes. *Proc. R. Soc. Lond.*
       A **159**, 496–506.   (SP II, 39)
    d The statistical theory of isotropic turbulence. *J. Aeronaut.*
       Sci. **4**, 311–15.   (SP II, 40)
    e The determination of drag by the Pitot transverse method.
       Rep. Memo. Aeronaut. Res. Comm. no. 1808.   (SP III, 21)
    f The determination of stresses by means of soap films. Article
       in *The mechanical properties of fluids*, pp. 237–54. London:
       Blackie.
1938a Plastic strain in metals. *J. Inst. Metals* **62**, 307–24.   (SP I, 27)
    b Analysis of plastic strain in a cubic crystal. *Timoshenko 60th*
       anniv. vol., pp. 218–24.   (SP I, 28)
    c Production and dissipation of vorticity in a turbulent fluid.
       Proc. R. Soc. London. A **164**, 15–23.   (SP II, 41)
    d The spectrum of turbulence. *Proc. R. Soc. Lond.* A **164**, 476–
       90.   (SP II, 42)
    e Measurements with a half-Pitot tube. *Proc. R. Soc. Lond.* A
       166, 476–81.   (SP IV, 13)
    f Turbulence. Chapter 5 of *Modern developments in fluid dynamics*,
       vol. I (ed. S. Goldstein), pp. 191–233. Oxford University
       Press.
    g Method of deducing $F(n)$ from the measurements.
       Appendix to paper by L.F.G. Simmons & C. Salter, *Proc. R.*
       Soc. Lond. A **165**, 87–9.
1939a (With A.E. Green) Stress systems in aeolotropic plates. I.
       Proc. R. Soc. Lond. A **173**, 162–72.   (SP I, 29)
    b Determination of the pressure inside a hollow body in
       which there are a number of holes communicating with
       variable pressures outside. Paper for Aeronaut. Res. Comm.   (SP III, 22)
    c The propagation and decay of blast waves. Paper for Civil
       Defence Res. Comm.   (SP III, 23)
    d Some recent developments in the study of turbulence. *Proc.*
       5th Int. Congr. Appl. Mech., Camb., Mass., 1938, pp. 294–310.
       Wiley.
1940a Propagation of earth waves from an explosion. Paper for
       Civil Defence Res. Comm.   (SP I, 30)
    b Notes on possible equipment and technique for experiments
       on icing on aircraft. *Rep. Memo. Aeronaut. Res. Comm.* no.
       2024.   (SP III, 24)
    c Generation of ripples by wind blowing over a viscous fluid.
       Paper for Chem. Defence Res. Dept.   (SP III, 25)
    d Notes on the dynamics of shock-waves from bare explosive
       charges. Paper for Civil Defence Res. Comm.   (SP III, 26)
    e Pressures on solid bodies near an explosion. Paper for Civil
       Defence Res. Comm.   (SP III, 27)

1940f The stagnation temperature in a wake. Paper for Aeronaut. Res. Comm. (SP III, 28)

1941a Calculation of stress distribution in an autofrettaged tube from measurements of stress rings. Paper for Advis. Coun. Sci. Res. Tech. Devel. (SP I, 31)

 b The propagation of blast waves over the ground. Paper for Civil Defence Res. Comm. (SP III, 29)

 c Analysis of the explosion of a long cylindrical bomb detonated at one end. Paper for Civil Defence Res. Comm. (SP III, 30)

 d The pressure and impulse of submarine explosion waves on plates. Paper for Civil Defence Res. Comm. (SP III, 31)

 e The formation of a blast wave by a very intense explosion. Paper for Civil Defence Res. Comm.

1942a The plastic wave in a wire extended by an impact load. Paper for Civil Defence Res. Comm. (SP I, 32)

 b (With R.M. Davies) The mechanical properties of cordite during impact stressing. Paper for Advis. Coun. Sci. Res. Tech. Devel. (SP I, 33)

 c The distortion under pressure of an elliptic diaphragm which is clamped along its edge. Paper for Co-ord. Comm. Shock-waves. (SP I, 34)

 d The dispersion of jets of metals of low melting point in water. Paper for Advis. Coun. Sci. Res. Tech. Devel. (SP III, 32)

 e The motion of a body in water when subjected to a sudden impulse. Paper for Co-ord. Comm. Shock-waves. (SP III, 33)

 f (With H. Jones) Note on the lateral expansion behind a detonation wave. Paper for Advis. Coun. Sci. Res. Tech. Devel. (SP III, 34)

 g (With R.M. Davies) The effect of the method of support in tests of damage to thin-walled structures by underwater explosions. Paper for Co-ord. Comm. Shock-waves. (SP III, 35)

 h The vertical motion of a spherical bubble and the pressure surrounding it. Paper for Co-ord. Comm. Shock-waves. (SP III, 36)

1943a (With R.M. Davies) The motion and shape of the shock induced by an explosion in a liquid. Paper for Co-ord. Comm. Shock-waves. (SP III, 37)

 b (With R.M. Davies) Experiments with 1/76th scale model of explosions near $\frac{1}{2}$-scale Asset target. Paper for Admiralty Undex Panel. (SP III, 38)

 c A formulation of Mr Tuck's conception of Munroe jets. Paper for Advis. Coun. Sci. Res. Tech. Devel. (SP III, 39)

 d (With H. Jones) Blast impulse and fragment velocities from cased charges. Paper for Advis. Coun. Sci. Res. Tech. Devel. (SP III, 40)

 e Note on the limiting ranges of large rockets. Paper for Ministry of Defence. (SP III, 41)

1944a (With H. Jones) The bursting of cylindrical cased charges.
Paper for Advis. Coun. Sci. Res. Tech. Devel.   (SP III, 42

  b Air resistance of a flat plate of very porous material. *Rep.
     Memo. Aeronaut. Res. Coun.* no. 2236.   (SP III, 43

  c The fragmentation of tubular bombs. Paper for Advis.
     Coun. Sci. Res. Tech. Devel.   (SP III, 44

  d (With R.M. Davies) The aerodynamics of porous sheets.
     Rep. Memo. Aeronaut. Res. Coun. no. 2237.   (SP III, 45

1945a (With A.E. Green) Stress systems in aeolotropic plates. III.
     Proc. R. Soc. Lond. A **184**, 181–95.   (SP I, 35)

  b Pitot pressures in moist air. *Rep. Memo. Aeronaut. Res. Coun.*
     no. 2248.   (SP III, 46

  c The genesis of the atomic bomb. I. Trying out the bomb *The
     Listener*, 16 August, pp. 173–174.

1946a The testing of materials at high rates of loading. *J. Instn Civ.
     Eng.* **26**, 486–518.   (SP I, 36)

  b The air wave surrounding an expanding sphere. *Proc. R. Soc.
     Lond.* A **186**, 273–92.   (SP III, 47

  c Note on R.A. Bagnold's empirical formula for the critical
     water motion corresponding with the first disturbance of
     grains on a flat surface. *Proc. R. Soc. Lond.* A **187**, 16–18. (SP IV, 14

  d Some model experiments in connection with mine warfare.
     Trans. Instn Nav. Archit., pp. 165–6.

1947a A connection between the criterion of yield and the strain
     ratio relationship in plastic solids. *Proc. R. Soc. Lond.* A **191**,
     441–6.   (SP I, 37)

1948a The formation and enlargement of a circular hole in a thin
     plastic sheet. *Q. J. Mech. Appl. Math.* **1**, 103–24.   (SP I, 38)

  b The use of flat-ended projectiles for determining dynamic
     yield stress. I. Theoretical considerations. *Proc. R. Soc. Lond.*
     A **194**, 289–99.   (SP I, 39)

  c The mechanics of swirl atomizers. *Proc. 7th Int. Congr. Appl.
     Mech. London, 1948*, part 1, 280–5.   (SP III, 48)

  d Eiver 1948. R. *Cruising Club J.*, pp. 194–200.

  e (With G. Birkhoff, D.P. MacDougall and E.M. Pugh)
     Explosives with lined cavities. *J. App. Phys.* **19**, 563–82.

1949a (With G.K. Batchelor) The effect of wire gauze on small
     disturbances in a uniform stream. *Q. J. Mech. Appl. Math.* **2**,
     1–26.   (SP III, 49)

  b The shape and acceleration of a drop in a high-speed air
     stream. Paper for Advis. Coun. Sci. Res. Tech. Devel.   (SP III, 50)

  c Aerodynamic properties of gauze screens. Appendix to *Rep.
     Memo. Aeronaut. Res. Coun.* no. 2276.

1950a   The dynamics of the combustion products behind plane and spherical detonation fronts in explosives. *Proc. R. Soc. Lond.* A **200**, 235–47.   (SP III, 51)

     b   (With R.M. Davies) The mechanics of large bubbles rising through extended liquids and through liquids in tubes. *Proc. R. Soc. Lond.* A **200**, 375–90.   (SP III, 52)

     c   The formation of a blast wave by a very intense explosion. I. Theoretical discussion. *Proc. R. Soc. Lond.* A **201**, 159–74.   (SP III, 53)

     d   The formation of a blast wave by a very intense explosion. II. The atomic explosion of 1945. *Proc. R. Soc. Lond.* A **201**, 175–86.   (SP III, 54)

     e   The boundary layer in the converging nozzle of a swirl atomizer. *Q. Jl Mech. Appl. Math.* **3**, 130–9.   (SP III, 55)

     f   The instability of liquid surfaces when accelerated in a direction perpendicular to their planes. I. *Proc. R. Soc. Lond.* A **201**, 192–6.   (SP III, 56)

     g   The path of a light fluid when released in a heavier fluid which is rotating. Paper for Aeronaut. Res. Coun.   (SP IV, 15)

     h   The 7th International Congress for Applied Mechanics. *Nature, Lond.* **165**, 258–60.

     i   Similarity solutions to problems involving gas flow and shock waves. *Proc. R. Soc. Lond.* A **204**, 8–9

1951a   Analysis of the swimming of microscopic organisms. *Proc. R. Soc. Lond.* A **209**, 447–61.   (SP IV, 16)

     b   The mechanism of eddy diffusivity. *Proc. General Discussion on Heat Transfer*, pp. 193–4. London: Institution of Mechanical Engineers.

1952a   Distribution of stress when a spherical compression pulse is reflected at a free surface. *Research* **5**, 508–9.   (SP I, 40)

     b   The action of waving cylindrical tails in propelling microscopic organisms. *Proc. R. Soc. Lond.* A **211**, 225–39.   (SP IV, 17)

     c   Analysis of the swimming of long and narrow animals. *Proc. R. Soc. Lond.* A **214**, 158–83.   (SP IV, 18)

     d   A scientist remembers. Hitchcock Lecture at the University of California (typescript only).

     e   Recollections of a scientist. Lecture given at some universities in Australia (typescript only).

1953a   Formation of a vortex ring by giving an impulse to a circular disk and then dissolving it away. *J. Appl. Phys.* **24**, 104.   (SP IV, 19)

     b   An experimental study of standing waves. *Proc. R. Soc. Lond.* A **218**, 44–59.   (SP IV, 20)

     c   Dispersion of soluble matter in solvent flowing slowly through a tube. *Proc. R. Soc. Lond.* A **219**, 186–203.   (SP IV, 21)

     d   Dispersion of salts injected into large pipes or the blood vessels of animals. *Appl. Mech. Rev.* **6**, 265–7.

1953e William Cecil Dampier 1867–1952. *Orbit. Not. Fell. R. Soc. Lond.* **9**, 55–63.

    f Rheology for mathematicians. *Proc. 2nd Int. Cong. on Rheology*, Oxford: Butterworth.

1954a The dispersion of matter in turbulent flow through a pipe. *Proc. R. Soc. Lond.* A **223**, 446–68.   (SP II, 43)

    b The use of a vertical air jet as a windscreen. *Mémoires sur la méchanique des fluides (Riabouchinsky Anniversary Volume)*, pp. 313–17.   (SP III, 57)

    c Conditions under which dispersion of a solute in a stream of solvent can be used to measure molecular diffusion. *Proc. R. Soc. Lond.* A **225**, 473–7.   (SP IV, 22)

    d The two coefficients of viscosity for a liquid containing air bubbles. *Proc. R. Soc. Lond.* A **226**, 34–9.   (SP IV, 23)

    e Diffusion and mass transport in tubes. *Proc. Phys. Soc. Lond.* B **67**, 857–69.

    f George Boole 1815–64. *Proc. R. Ir. Acad.*, pp. 66–73.

1955a The action of a surface current used as a breakwater. *Proc. R. Soc. Lond.* A **231**, 466–78.   (SP IV, 24)

1956a Strains in crystalline aggregates. *Proc. Colloq. Deformation and Flow of Solids, Madrid, 1955*, pp. 3–12. Berlin: Springer.   (SP I, 41)

    b (With J.C.P. Miller) Fluid flow between porous rollers. *Q. Jl Mech. Appl. Math.* **9**, 129–35.   (SP IV, 25)

    c Fluid flow in regions bounded by porous surfaces. *Proc. R. Soc. Lond.* A **234**, 456–75.   (SP IV, 26)

    d (With B.R. Morton & J.S. Turner) Turbulent gravitational convection from maintained and instantaneous sources. *Proc. R. Soc. Lond.* A **234**, 1–23.   (SP II, 44)

    e George Boole, F.R.S. 1815–64. *Notes Rec. R. Soc. Lond.* **12**, 44–52

    f An applied mathematician's apology. Address of reply after receiving the de Morgan Medal from the London Mathematical Society. See Appendix A herein.

1957a (With P.G. Saffman) Effects of compressibility at low Reynolds number. *J. Aeronaut. Sci.* **24**, 553–62.   (SP III, 58)

    b Fluid dynamics in a papermaking machine. *Proc. R. Soc. Lond.* A **242**, 1–15.   (SP IV, 27)

1958a Flow induced by jets. *J. Aero/Space Sci.* **25**, 464–5.   (SP II, 45)

    b (With P.G. Saffman) The penetration of a fluid into a porous medium or Hele–Shaw cell containing a more viscous liquid. *Proc. R. Soc. Lond.* A **245**, 312–29.   (SP IV, 28)

1959a (With P.G. Saffman) A note on the motion of bubbles in a Hele–Shaw cell and porous medium. *Q. J. Mech. Appl. Math.* **12**, 265–79.   (SP IV, 29)

1959b The dynamics of thin sheets of fluid. I. Water bells. *Proc. R. Soc. Lond.* A **253**, 289–95. (SP IV, 30)

c The dynamics of thin sheets of fluid. II. Waves on fluid sheets. *Proc. R. Soc. Lond.* A **253**, 296–312. (SP IV, 31)

d The dynamics of thin sheets of fluid. III. Disintegration of fluid sheets. *Proc. R. Soc. Lond.* A **253**, 313–21. (SP IV, 32)

e (With P.G. Saffman) Cavity flows of viscous liquids in narrow spaces. *Proc. 2nd Symp. on Naval Hydrodynamics, Wash., 1958*, pp. 277–91. Washington: Office of Naval Research.

f The present position in the theory of turbulent diffusion. Article in *Proc. Symp. on Atmospheric Diffusion and Air Pollution*, pp. 101–12. Academic Press.

g Obituary of R.M. Davies. *Phys. Soc. Year Book, 15*.

1960a Formation of thin flat sheets of water. *Proc. R. Soc. Lond.* A **259**, 1–17. (SP IV, 33)

b Deposition of a viscous fluid on a plane surface. *J. Fluid Mech.* **9**, 218–24. (SP IV, 34)

c Similarity solutions of hydrodynamic problems. Article in *Aeronautics and Astronautics (Durand anniv. vol.)*, pp. 21–8. Pergamon.

1961a Deposition of a viscous fluid on the wall of a tube. *J. Fluid Mech.* **10**, 161–5. (SP IV, 35)

b Interfaces between viscous fluids in narrow passages. Article in *Problems of Continuum Mechanics (Muskhelishvili anniv. vol.)*, pp. 546–55. Philadelphia: Soc. Indus. & Appl. Math.

c Fire under influence of natural convection. Publication 786, Nat. Acad. Sci.—Nat. Res. Coun., Washington, pp. 11–31.

1962a On scraping viscous liquid from a plane surface. *Miszellaneen der angewandten Mechanik (Festschrift Walter Tollmien)*, pp. 313–15. (SP IV, 36)

b Standing waves on a contracting or expanding current. *J. Fluid Mech.* **13**, 182–92. (SP IV, 37)

c Gilbert Thomas Walker 1868–1958. *Biogr. Mem. Fellows R. Soc. Lond.* **8**, 167–74.

d Review of *Hydrodynamic and Hydromagnetic Stability*, by S. Chandrasekhar, Oxford Univ. Press 1961. *Proc. Phys. Soc.*

1963a Cavitation of a viscous fluid in narrow passages. *J. Fluid Mech.* **16**, 595–619. (SP IV, 38)

b Memories of Kármán. *J. Fluid Mech.* **16**, 478–80.

c Scientific diversions. Article in *Man, Science, Learning and Education* (ed. S.W. Higginbotham), pp. 137–48. Rice Univ. Semicent. Publ.

d Sir Charles Darwin (1887–1962). *Am. Phil. Soc. Y.*, pp. 135–40.

1964a   Disintegration of water drops in an electric field. *Proc. R. Soc. Lond.* A **280**, 383–97.   (SP IV, 39)

    b   Cavitation in hydrodynamic lubrication. Article in *Cavitation in Real Liquids* (ed. R. Davies), pp. 80–101. Elsevier.

    c   The life of George Boole. Address at the Boole Centenary Celebrations, Lincoln (typescript only).

    d   Review of *Fluidised Particles*, by J.F. Davidson & D. Harrison, Cambridge Univ. Press, 1963.

1965a   (With A.D. McEwan) The stability of a horizontal fluid interface in a vertical electric field. *J. Fluid Mech.* **22**, 1–15.   (SP IV, 40)

    b   Note on the early stages of dislocation theory. Article in *Sorby Centennial Symposium on the History of Metallurgy (ed. C.S. Smith), pp. 355–8. Gordon & Breach.*

1966a   Conical free surfaces and fluid interfaces. *Proc. 11th Int. Congr. Appl. Mech., Munich, 1964,* pp. 790–6.   (SP IV, 41)

    b   The force exerted by an electric field on a long cylindrical conductor. *Proc. R. Soc. Lond.* A **291**, 145–58.   (SP IV, 42)

    c   The circulation produced in a drop by an electric field. *Proc. R. Soc. Lond.* A **291**, 159–66.   (SP IV, 43)

    d   Oblique impact of a jet on a plane surface. *Phil. Trans. R. Soc. Lond.* A **260**, 96–100.   (SP IV, 44)

    e   (With A.D. McEwan) The peeling of a flexible strip attached by a viscous adhesive. *J. Fluid Mech.* **26**, 1–15.   (SP IV, 45)

    f   When aeronautical science was young. *J. Aeronaut. Soc.* **70**, 108–13.

    g   Motion of solid bodies in rotating fluids. Typescript of paper read at IUTAM Symp. Rotating Fluid Systems, La Jolla, March 1966.

1967a   Low-Reynolds-number flows. 16 mm. colour sound film, produced by Educational Services Inc.

1968a   The coalescence of closely spaced drops when they are at different electric potentials. *Proc. R. Soc. Lond.* A **306**, 423–34.   (SP IV, 46)

    b   Review of *An Introduction to Fluid Dynamics*, by G.K. Batchelor, Cambridge Univ. Press, 1967. *Eureka.*

1969a   Instability of jets, threads and sheets of viscous fluid. *Proc. 12th Int. Congr. Appl. Mech., Stanford, 1968,* pp. 382–8.   (SP IV, 47)

    b   Motion of axisymmetric bodies in viscous fluids. Article in *Problems of Hydrodynamics and Continuum Mechanics (Sedov anniv. vol.),* pp. 718–24. Philadelphia: Soc. Indus. & Appl. Math.   (SP IV, 48)

    c   Electrically driven jets. *Proc. R. Soc. Lond.* A **313**, 453–75.   (SP IV, 49)

    d   Amateur scientists. *Michigan Q. Rev.* **8**, 107–13.

1969e   Aeronautics fifty years ago. *Quest: J. City Univ.*, no. 8, pp. 12–19.

    f   (With J.R. Melcher) Electrohydrodynamics: a review of the role of interfacial shear stresses. *Ann. Rev. Fluid Mech.* **1**, 111–46.

1970a   The interaction between experiment and theory in fluid mechanics. *Bull. Brit. Hydromech. Res. Ass.*; and *Ann. Rev. Fluid Mech.* **6** (1974), 1–16.

    b   Some early ideas about turbulence. *J. Fluid Mech.* **41**, 3–11.

1971a   A model for the boundary condition of a porous material. I. *J. Fluid Mech.* **49**, 319–26.

    b   Aeronautics before 1919. *Nature, Lond.* **233**, 527–9; and *Bull. Inst. Math. Applic.* **10** (1974), 363–6.

    c   The history of an invention. *Eureka* **34**, 3–6; and *Bull. Inst. Math. Applic.* **10** (1974), 367–8.

    d   Aeronautical experience before 1919. Lester Gardner Lecture at Mass. Inst. Technology (Typescript only).

1973a   (With D.H. Michael) On making holes in a sheet of fluid. *J. Fluid Mech.* **58**, 625–39.

    b   Memories of von Kármán. *Soc. Indus. Appl. Math. Rev.* **15**, 447–52.

    c   The stability of a conducting jet in an electric field. *Proc. 10th Symp. on Advanced Problems and Methods in Fluid Mechanics, Poland,* 1971, pp. 9–16. Polish Academy of Sciences.

Index

Adams Prize for 1915, 120, 158–9
aeronautics, birth of, 63 *et seq.*
agreement between theory and
    experiment, 60–2, 250
airship crash, 69
airship trip, 181
amateur, G.I. as a scientific, 224–5
Archimedes principle, 111, 262
Atlantic, flight across the, 75–7
atomic bomb, 202–7
    test of, 208–10

balloon trip, 74
Barenblatt, G.I., 249
Barnes, H.T., 54
Bell, E.T., 9, 10
Benjamin, T. Brooke, 227
blast waves
    in air, 192–6
    in water, 197–200
bombs on Cambridge, 191
Boole,
    Alice (m. Stott), 18, 28
    Ethel Lillian (m. Voynich), 10, 20,
        28
    George, 5 *et seq.*
    John, 5, 6, 8
    Joshua, 6, 8
    Lucy, 20, 28, 30
    Margaret (m. Taylor), 16, 25
    Mary (m. Hinton), 16
    Mary Everest (m. Boole), 6, 11, 15
        et seq.
Brooke, W., 9
bubble harbour, 200–1
bubble rising through water, 215

Burgers, J.M., 175, 181–3
Busk, E., 64–5

Cambridge Mathematical Journal, 13, 31
Carpenter, H.C.H., 144
Cavendish Laboratory, 40, 41, 117
Cherry, T.M., 219
Christmas lectures at Royal
        Institution,
    on wireless telegraphy, 30
    on ships, 111
Chudleigh, 64
Congress of Applied Mechanics,
    1st International, Delft 1924, 165, 178
    2nd International, Zürich 1926, 95
    3rd International, Stockholm
        1930, 162
    4th International, Cambridge, UK
        1934, 185
    5th International, Cambridge,
        Mass. 1938, 186
    13th International, Moscow 1972, 246
correspondence between
    Burgers and Taylor, 181–3
    Prandtl and Taylor, 183–8
C.Q.R. anchor, 112–16

DAMTP, 226
darts, 65
Davies, Gladys, 248
Davies, R.M., 199, 215
D-Day, 201–2
deformation and break-up of drops,
    125–8
diffusion by continuous movements,
    162–4

282